Grape Whisperers

Allen G. Holstein

ISBN: 978-1-66786-631-4

Introduction

IF YOU ARE HOLDING THIS book in your hands, the odds are that you are visiting one of the Willamette Valley's hundreds of wineries, tasting rooms, or restaurants. Or perhaps you live, work, or study in one of the other wine regions of the world and wonder how the Oregon miracle happened.

Whoever and wherever you are, you're wondering, how did the entire world beat a path to these rolling hills? We started from next to nothing. How is it that top French, Australian, Californian (and most recently, Italian) wine producers established or purchased major vineyards in an area once known for hazelnut and walnut orchards and Christmas tree farms? How is it that Oregon's Pinot Noir and other wines have taken their place among the world's finest?

Clearly, they have. It's not just local hubris. Oregon accounts for about 1.5 percent of the volume of US wine sold annually, but represents 17 percent of the American wines rated at 90 or higher by *Wine Spectator* magazine, which is the industry's bible. It is a hundred-point scale. Oregon is clearly punching above its weight. Oregon wine is also sold internationally, in Canada, Mexico, Britain, Europe, Hong Kong, Japan, and China.

The Oregon wine experience for me started, perhaps ironically, in Kentucky. I was getting a master's degree in horticulture at the University of Kentucky. My professor and mentor there, Carl Chaplin, had a son on the faculty at Oregon State University. Carl said, "I think you would like Oregon. You ought to apply for graduate school there. I will fix you up with my son." Carl knew I had a love of the land and the outdoors. This was 1979. Talk about serendipity.

That led to my getting hired the following year as a manager of Knudsen Vineyards in Dundee. It supplied to the Knudsen-Erath Winery, one of the thirteen bonded wineries that existed at the time. I made five dollars an hour and lived in a small cabin on a hill. I thought it would be a two- or three-year stint. Instead, it turned out to be a career plunge and a multi-generational leap in the sense that my son, Jackson, and his wife, Ayla, have continued the family tradition by establishing the Granville Wine Company using land that I purchased in 1984.

The vast majority of the people who were involved in the wine industry in the beginning came from outside of Oregon. Few dared to quit their day jobs because no one knew what the future might hold. It might have been that we were all crazy dreamers pouring money into a giant sinkhole with no future. The late Cal Knudsen, who emerged as my mentor, was from Seattle and had been the chief executive officer of MacMillan Bloedel, a Canadian timber company. He was a visionary leader and bought acreage in the Dundee area starting in 1971 because the rolling hills reminded him of the Champagne region of France. Dick Erath was an electrical engineer from the San Francisco Bay area beginning a second career with a young family. The two men had a partnership in Knudsen Erath Winery, with both owning their own vineyards.

What helped all of us was that the soils and climate of the Willamette Valley are uniquely fertile. That's what attracted entire wagon trains of settlers to cross the Rocky Mountains and settle in the valley in the 1840s and 1850s. They were attracted to the rich soils of the valley floor, but the wine industry discovered that the hills were more austere and hence were better suited to growing fine grapes. That's why grapes planted here, on roughly the same latitude as Western Europe and ten hours by car north of San Francisco, are so excellent. We also absorbed ideas about how to trellis and manage our vineyards from France, California, and Australia. The area's openness to ideas, not just money, from all over the world has been one of its key strengths. There was a process of cross-pollination. Or in wine terms, you might call it cross-fermentation.

On the way, we had to overcome the dominance of the California wine industry. We had to figure out new and different ways to sell our wines because the big national liquor and wine distributors want to do business primarily with large-scale wineries that can produce thousands of cases of a particular wine for wide distribution to liquor and wine stores. Although a handful of big Oregon wineries can do that now, the vast majority of Oregon's wineries are smaller and niche-oriented. They tend to make the truly excellent wines. But they cannot satisfy the demands of the big distributors.

So we gradually found ways to build a wine tourism industry that has attracted millions of visitors, allowing smaller producers to sell much of their wine directly to guests. Groups of wine enthusiasts come to Oregon on wine tours from all over the United States to buy the vintages they love most as they gaze out over our hills. Airbnb transformed the landscape by making so many more homes and rooms available for wine enthusiasts. Wine clubs on the Internet have been another niche way we have found to sell our wines. It took a ruling from the US Supreme Court in 2005 to overturn laws in New York and Michigan preventing out-of-state wineries from selling in those states. That, and improvements in packaging, allowed us to start shipping wine directly to customers all over the country.

As an industry, we've had to survive a great many hardships. The eruption of Mount St. Helens in 1980 covered the region in ash. The phylloxera insect infestation, which spread invisibly underground from root to root, threatened to wipe us out in 1990. We've had to contend with other exotic pests as more and more vineyards were planted. Even before the recent conversation about climate change, we suffered weather events such as harsh winter freezes. The recent burst of wildfires across the American West ruined part of our crop in 2020, at the same time that we were contending with the coronavirus pandemic. Oregon's grape production was down 29 percent as a result, according to the Oregon Wine Board. Chronic wildfires, extreme heat, and shortages of water are challenging all agriculture in the American West, and we are not immune from those pressures. Temperatures hit a record 114 degrees Fahrenheit prior to the 2021 harvest. We also are highly dependent on seasonal laborers from Mexico. So our fate is closely connected to the swirling debates about immigration.

In the pages that follow, my co-participants and I will tell the story of how the Oregon wine scene developed in terms of "waves." I personally specialized in the details of horticulture—how many vines should be planted per acre and how, what techniques should be used to manage the grapes, and how to estimate the total volume of grapes that would be harvested. I personally planted almost a thousand acres of grapes. But not many vineyard managers actually make their own wine. That task typically falls to winemakers, who possess very different skill sets. Reasonable people may differ over which set of personalities—vineyard managers or winemakers—possess the greatest egos.

To round out the full story of Oregon's wine boom, I have turned to some of those winemakers and other key personalities to add their voices. One of them is Ken Wright, the founder originally of Panther Creek Cellars and ultimately of Ken Wright Cellars, which has become an iconic brand. Ken went to high school with me in Louisville, Kentucky, and we were roommates for a while at the University of Kentucky. Ken developed a taste for fine wine while working at an upscale French restaurant in Lexington. Our interest in wine led us to Oregon at roughly the same time.

Another voice is that of Véronique Drouhin (now married as Véronique Boss), who led her family's initiative to launch a vineyard and winery in Oregon. Oregon's connection with Burgundy was critically important because the Pinot Noir grape comes from Burgundy.

You will also hear from the late Cal Knudsen, founder of Knudsen Vineyards; Susan Sokol Blosser, one of the original thirteen pioneering wine families; and Rollin Soles, who was the winemaker at Argyle and recently sold a majority of his own winery to a major Italian winemaker.

We all had a hand in putting Oregon on the world's map of fine wine regions. It does not happen very often that a major wine district is born. It happened over a period of fifty years, which is a very short time comparatively.

Think how many centuries it has taken the Burgundy or Champagne regions of France to evolve. They've been making wine since the time of the Romans. The Oregon wine explosion has been a uniquely American story of innovation and risk-taking.

Table of Contents

CHAPTER ONE:

Wandering in the Wilderness,

1975–1980

WHEN KEN AND I WERE coming of age in the 70s, there were two mega-trends that shaped our thinking. The first Earth Day in 1970 helped usher in a budding environmental movement. As part of that, my father, Bill Sr., was very active in the Sierra Club. An avid outdoorsman, he also was very active politically. That allowed him to meet prominent, like-minded folks, one of whom was Wendell Berry, a famous Kentucky author and poet. He was a major back-to-the-land proponent.

For what felt like a final farewell hike just before leaving for freshman year of college at the University of Kentucky, my dad organized a backpacking trip to Yellowstone for me, Ken, and my brother in September 1972. The only problem was we lived in Kentucky. So we decided to take turns driving for thirty hours, stopping only for gas and bathroom breaks. Then when we arrived, we intended to put on our packs and charge off to the woods. The idea was so then uncommon that the local paper, the

Louisville Courier-Journal, sent along a photographer to document what these crazies were doing. The experience was profound. We spent a week out in the wilderness and did not see another human being.

When Ken Wright was reviewing this book manuscript in 2022, the Yellowstone trip came up in conversation. We both looked at each other fifty years later and agreed that, independently, we both walked out of Yellowstone's Lamar Valley that September saying to ourselves, "I want to work with nature."

Wendell Berry was also a member of the Sierra Club, which organized weekend hiking and canoe trips in the Red River Gorge of Kentucky. I went on many hiking trips there with my dad, Wendell, and other personalities. I sat around the campfire in my teens listening to their stories.

One time, before the advent of email, my dad had some papers he wanted Wendell to have. Ken and I drove up from Louisville to his farm. He was expecting us, but he was plowing a field behind his horse and could not be bothered. He waved us up to his house where we delivered the papers to his daughter who invited us to shuck corn with her. It was the real deal. We saw Wendell Berry's farm.

Berry was also a professor at the University of Kentucky. One literature assignment I had was to review all the works of my favorite author and summarize their work. I did better than that and went up to campus to interview him after having read all of his books. I had two hours one on one with him. This was some years after sitting around the fire with him. He obviously had a big influence on me. He preached about local agricultural economies and environmental impacts. I was never as pure as him in practice, but I carried those ideas with me. He was passionate about having a "sense of place." That's what I felt when I arrived in Dundee. And it was what Ken felt when he landed in Carlton. These were not just places we lived. We belonged.

The second megatrend in this rough time frame was the coming of age of Napa Valley. Their winemakers had recently won a tasting competition in France, and it opened the eyes of many Americans to the possibility of actually growing wine grapes in the United States. The growing success of Napa caused "back to the land" types in many other states to wonder if they could do it too, even in Kentucky. I was in the University of Kentucky's School of Agriculture and learned that the state was desperate to find an alternative to tobacco as an agricultural enterprise. Ken, who was a housemate for three years, and I did some research and learned that Americans drank four liters per capita of wine per year and the French drank hundred. We saw a great upside.

My master's degree work was funded by a coal company. They did not like the bad press associated with strip mining. They wanted to plant a vineyard on a strip mine spoil to prove they were not raping the earth. So I drove two hours out from Lexington to deep dark Appalachia to a strip mine near the town of Jackson, Ky. It turns out Jackson would much later become one of the settings for a book called *Hillbilly Elegy* by J. D. Vance. I planted a one-acre postage-stamp size vineyard on the strip mine site. What had once been rolling hills was now flat as a pool table. I worked hard on it. The grapes grew just fine. But I don't think the coal company ever got the kind of positive publicity they had hoped for. The people planting vineyards in Kentucky, including me, never thought about how and where to actually sell the wine. We had no idea what to actually do with the grapes.

Like all college kids, Ken and I smoked dope and drank cheap Portuguese wine. Along the way, Ken got a job at a nice restaurant in Lexington, which defies the traditional stereotype of Kentucky. It is a college town surrounded by horse country with many wealthy and cosmopolitan individuals present from different countries. The restaurant was called the Fig Tree. As part of his job, Ken got to try some nice wines being sold by the restaurant. It opened his eyes. Most were French. He started bringing some home and we asked each other, "How can two wines be so different?" We didn't know it at the time, of course, but we were starting on the journey of our lifetimes.

The university had a noncredit division where anybody could teach anything to anyone if they could design a course and get people to sign up for it. Ken and I had an idea to teach a wine appreciation class. We had been trying random wines from the restaurant and wanted to learn more but certainly did not have the expertise to teach. We got people to sign up, and we sweet-talked them into hosting classes in their homes. Most who signed up were older than we were, some of them faculty members. All we did was go to the bookstore and buy a book: *Wine Regions of the World*. Before each class, we would read it and then go repeat what we had learned. In the process, we learned about the Appellation d'Origine Contrôlée of Burgundy and tried some of the best Burgundies on someone else's nickel. We were hooked. For our personal use, we even learned a trick. We would go to a liquor store, pick out an expensive Burgundy, and then exchange the price tag with that of a low-end Italian wine. The clerks never knew any better. We were hooked on Burgundies.

We were enthralled with the idea of starting a winery in Kentucky. Ken's dad was a marketing executive with Brown Forman, a major spirits company based in Louisville. That's maybe why Ken had marketing in his blood. Brown Forman had recently purchased Bolla Soave, an Italian white wine company. Ken and I thought they might have a vision for the future that included fine wine. So we came up with a plan to have them finance our winery in Kentucky. My dad was an accountant, and he helped us come up with a reasonable cash flow projection for the project. Numbers were in my blood.

We talked Ken's dad into getting us an audience with the chief executive officer and senior executives of Brown Forman at our tender age of twenty-two. We got all dressed up, but it snowed on the day of the meeting, and Ken had to push my pickup truck through the snow so we could get to the plowed parking area. He managed to keep the mud off his sport coat.

We got into the conference room with the executives and began to explain how we were going to sell twenty thousand cases of different wines out of a tasting room, which these guys knew was impossible. This was Kentucky. Ken's dad sat in the back of the room and waved us off, telling us to kill the presentation before we finished it. The CEO explained they needed fifty thousand cases of one wine label to justify a national marketing campaign. We were completely out of our league.

CHAPTER TWO:

The Pioneer Struggle Years,

1980–1985

AFTER OUR HUMILIATION AT BROWN Forman, Ken moved to the University of California at Davis to take up his studies. I later visited him there, thinking of entering a PhD graduate program. But instead, I enrolled at Oregon State University in the fall of 1979. Oregon had a fledgling wine industry. I bought some Oregon Pinot Noir and thought it had the fine balance of some of the Burgundies we had tried in Kentucky.

During the Christmas break of 1979, I decided to stop by the Knudsen Erath Winery to taste their wines. It was the largest winery by far with sixty-six acres of grapes already planted on site and another thirty acres at another location. I met Kina Erath, then wife of Dick, who was running the tasting room. I explained that I had an interest, and she responded by saying they were going to hire a vineyard foreman. I applied and was eventually offered the job. I ended my academic career and started work in March 1980. I was one of the first "hired guns." Ken initially took a job at a winery in Monterey, California.

When we got started in the early 1980s, no one could have imagined that the world's giants would one day knock on our door. There were thirteen wineries in Oregon at that time. (See table.) I was one of the first hired guns. I attended social events where all thirteen were together. They were convivial affairs. Everyone seemed to recognize that if one grower started distributing a poor quality wine, it would reflect on the reputation of Oregon wines in general, and everyone would suffer.

THE ORIGINAL 13 WINERIES

1. Hillcrest, led by Richard Sommers, was the first winery in Oregon, but it was in southern Oregon. He was a go-with-the-flow guy, preferring to leave it all to Mother Nature.

2. Eyrie was first in the Willamette Valley region of Oregon to plant Pinot Noir under the leadership of David Lett. His 1975 Pinot Noir placed well in a French competition, and it put Oregon on the map.

3. Knudsen and Erath were second in the Willamette Valley and were the largest with sixty-six acres of grapes around the winery and another thirty acres nearby.

4. Honeywood made mostly fruit wine.

5. Sokol Blosser was led by Bill and Susan Sokol Blosser. Bill was a city planning professor and was instrumental in developing Oregon's land use laws that protected vineyard land from McMansions.

6. Adelsheim was begun by Dave and Ginny Adelsheim. I sold them grapes from my vineyard for a time. Dave has been a tireless champion of Oregon's potential for decades.

7. Ponzi Winery was another family affair that stayed in the family until 2021.

8. Elk Cove Winery was started by Pat and Joe Campbell. Joe was an emergency room doctor, and Pat ran the vineyard operation. Their son, Adam, runs the operation today.

9. Amity was started by none other than Myron Redford, who started planting in the early 1970s and is considered a father of the industry.

10. Oak Knoll were famous for their "Bachus Goes Bluegrass" concerts in the 1980s.

11. Tualatin was one of the three largest wineries in Oregon.

12. Martha Maresh and Fred Arterberry. Martha's father, Jim, was the patriarch of the Worden Hill Road circuit, selling off parcels of land to like-minded folk such as Art and Vivian Weber. Fred's parents funded a winery specializing in sparkling wine and sourcing grapes from the Maresh Vineyard.

13. Cote Coloumbe was located near Banks and did not survive the challenges of the 1980s.

Notes: Of the original thirteen wineries, only Eyrie, Sokol Blosser, and Elk Cove are still owned by the founding families.

Notes: There were other groups that had vineyards but not wineries, which actually make the wine. They were Ken and Penny Durant, who had vineyards and later started a winery and olive oil mill. Bethel Heights had planted a vineyard and would later make wine. The next generation of Maresh began to make wine. The same is true of Knudsen Vineyards. All are still in family hands. To my knowledge, these four vineyards, combined with the Eyrie, Sokol Blosser, and Elk Cove wineries, are the only businesses that have been in the game longer than Holstein Vineyards.

The French influence on the pioneers was clear. They knew that the Willamette Valley's climate resembled that of the Burgundy region of France, where Pinot Noir grapes originated. Bill Blosser had spent a year in France during college, taking classes in Paris and working as a dishwasher at a resort near Grenoble, according to Susan Sokol Blosser's book *The Vineyard Years: A Memoir With Recipes.* "He had emerged from that experience a wine-drinking, French-speaking Francophile, with a worldliness that attracted me to him in college," she wrote. "I had spent the summer after high school studying in France, and I had grown up drinking wine from my father's wonderful wine cellar."

She added: The early pioneers "stood out with their quirky individuality. Scruffy sideburns, beards and mustaches aside, they were smart and enterprising . . . With diverse backgrounds in engineering, music, philosophy, history, and the humanities, coupled with a fierce spirit of independence, we were united in a passion for Pinot Noir." Bill Blosser often wore a beret as he worked in his vineyard.

All the pioneers planted against all advice from the California Wine Establishment. There was no bank financing available because it was not an established industry with a track record of success. Banks and other lending institutions did not trust it. As a result, the first plantings were small-acreage and decidedly modest. The pioneers simply put some own-rooted vines in the ground. Only years later did they put in trellising when the growth of the vines demanded it and cash flow allowed it. They knew that California was planted with phylloxera-resistant grafted vines, but grafted vines cost more. And no one knew whether Oregon would ever have to worry about phylloxera. (A rootstock can be "grafted" with a different clone of grape to give it slightly different characteristics, such as insect resistance. Grafted vines are different from own-rooted vines. Grafted vines have Vitis vinifera grafted onto insect-resistant rootstocks of a different Vitis species that is native to the Eastern United States. See Glossary, page 32.)

The first plantings were small. The pioneers followed with additional plantings in subsequent years of a size that current cash flow would allow. Consequently, the first wines that flowed from those plantings were low in volume. The wines showed promise, but there was not enough volume for them to reach deeply into the consumer market.

Much later, when interviewed, Bill Blosser was asked what his goals were during that time. He said it was to "just make through the year." That was what it was like. The locals never thought we crazy out-of-staters would succeed. I remember the day an insurance agent rolled up to the tasting room at Knudsen Erath Winery. It was closed, and I was the only guy he could find, and he started pitching me on insurance. I wanted to kiss him because it was the first time anyone had thought of us as a business.

I started for five dollars an hour plus housing at the cabin, which was located on Cal Knudsen's land. I went down the first day at 7 a.m. because Erath had told me that was when the Mexican workers showed up. I expected that Erath would join us and explain that I was the new boss. He didn't. In fact, I never really got a walk around with him. I found out after I started that Cal Knudsen owned the sixty-six acres of vineyard surrounding the winery, which Erath co-owned in a fifty-fifty percent partnership with Knudsen. Erath had a remote vineyard of thirty acres seven miles away. I had to drive a tractor on a country road to get there. The two men co-employed me in their respective vineyards. There were altogether a thousand acres of vineyards in the state at the time and I, with my vast knowledge, was in charge of a hundred of them. I soon wished I had studied Spanish and mechanics instead of horticulture.

At the time I started, Erath had just returned from a trip to Burgundy, another example of that region's impact on us. He described a system of vertical trellising that did not exist to my knowledge in America. In it, there was a system of nails installed in the line posts on which wires were manipulated to force the vines to grow vertically. We began retro-fitting Knudsen Vineyards in 1980. It was the first in Oregon. The system was gradually adopted regionally as the trend toward tighter row spacing required it. It also helped with disease control. It is now standard practice on the West Coast and the US version of vertical shoot positioning started in Oregon.

Seen in retrospect, the pioneers had good instincts when it came to site selection. They knew from their travels that the better wines in Europe came from hillsides and not the valley floor. So they planted mostly on hillsides. That was fortunate because Napa at the time had planted many acres on the valley floor. The soil on the side of a hill tends to be less rich, because of Oregon's unique geology, than the soil of the valley floor. Less rich soil makes the vines struggle. It's generally believed that making vines struggle improves the taste of the grapes by making them more distinctive and improving the quality of the wine as a result.

The amount of clay in the soil is one key. Robert Drouhin, the famous French winemaker from Burgundy who would much later make a critical investment in Oregon, once explained to me that he was on a committee in France that was reexamining the boundaries of French appellations, or wine districts. The French were puzzled about why the same grape variety planted in two vineyards just meters apart could produce such different tastes. "It is something about the clay, All-annn," he said. "We don't know what it is, but it is about the clay."

That bounced around my brain for a while until I reconciled it with my university training in soils. Jory is the name of the soil in the Dundee Hills. It has high clay content. "Field capacity" is the amount of water in a soil when it can hold no more water. Jory has 30 percent water by weight at field capacity. Any additional rainfall runs off or evaporates. "Wilting point" is the amount of water in the soil when plants can no longer extract water from it. Jory still has 20 percent moisture content at its wilting point. Clay competes with plants for moisture at the dry end of the soil moisture spectrum. So the vines have to fight for it. It is not that way in sandy or silty soils.

The pioneers experimented to figure out which varieties of grapes were best suited to Oregon's cooler climate. They had a passion for Pinot Noir and Chardonnay, two of the six wines that Europeans considered to be "noble varieties" worthy of being consumed by a king. A fair bit of Riesling also was planted, a third noble variety. All the pioneers had a little block of what was jokingly referred to as "chateau fruit salad," with ten vines of every obscure variety known to humankind. It was like they had a meeting before I arrived and asked one another, "What if Pinot and Chardonnay don't work?" So they kept searching for their magic elixirs.

Erath planted Gewurztraminer, Sauvignon Blanc, and Merlot in his vineyard. Nobody bet 100 percent on Pinot Noir. I came to possess the original vineyard maps of Dick Erath and David Lett and could see how much of what they planted and when. It was modest. They clearly were experimenting.

But ultimately it was their passion for Pinot Noir that prevailed. "They decided this was Pinot Noir country," Ken Wright says. "We were so fortunate that the earliest producers decided this region was meant for Pinot Noir, and they were right. All of us owe them a tremendous debt."

The Pinot Noir grape originated in Burgundy, but the Burgundians did not, and still do not, call their wines "Pinot Noirs." They name them for the places they are grown. This is the "terroir" system in which a specific wine comes from a specific piece of land. Take a Volnay or Pommard. Although they are 100 percent Pinot Noir, they are called a Volnay or Pommard. When Oregon started to make and sell Pinot Noirs, one reason it was difficult to build recognition was that there were so few Pinot Noirs on the market. Oregon could not ride on the fame of Pinot Noir.

Even after Ken and I arrived, the early vineyard owners and winemakers did not know what clones of the specific types of grapes should be planted. Just as identical human twins have the same DNA but have slightly different characteristics, two clones of the same grape strain present different characteristics. There was an awareness among some of the pioneers of a whole universe of clones of Pinot Noir and Chardonnay that existed in France and that did not exist in America. David Adelsheim of Adelsheim Vineyards was prominent in an effort to import clones, but that would take years. Owing to the US Department of Agriculture rules, we could not buy plants from France, so we were left with the selection available in California. But generally it was a time before clonal awareness in the US.

There was just one clone of California Chardonnay, called the Davis 108 clone, that was available to us in Oregon. It was unfortunate that all of the Chardonnay planted in Oregon in the 1970s and 1980s was this clone. It was the clone responsible for much of the success of California Chardonnay. But in Oregon, it was a struggle to get it to ripen because our growing season was cooler than California's.

Once I was walking the rows of these Davis 108 clones in Oregon with Robert Drouhin. He mused out loud, "I wonder if it is really Chardonnay?"

We started debating. I asked him three questions about this grape's behavior relative to Pinot Noir grapes in Burgundy. Did Chardonnay clusters in Burgundy always weigh more than Pinot Noir clusters? Did Chardonnay in Burgundy always bloom later than Pinot Noir? Did Chardonnay in Burgundy always ripen later than Pinot Noir? The answer to all three questions in Oregon was "yes," but Drouhin told me the answer in Burgundy was "no."

Fast forward almost twenty years, and the science had improved to the point that it could determine that the original clone 108 had genetic material from Pinot Noir but also from a lesser known variety Gouais Blanc. That meant that all of the famous California Chardonnay produced over the decades came from an imposter clone. The early Californians chose it because it produced more grapes—and Robert Drouhin could tell by walking the rows. Oregon eventually brought in genuine Chardonnay clones.

There were two clones of Pinot Noir, named after villages in Switzerland and Burgundy, Wadensville and Pommard. They were good clones and responsible for the early successes in Oregon.

A broader selection of both Pinot Noir and Chardonnay served like more colors on an artist's palette. They created more blending opportunities and less monolithic wines.

Even if the pioneers had been willing to foot the cost of phylloxera-resistant rootstocks, nobody had any idea of which one or ones to use. Oregon's soil was vastly different from California's. So that justified the use of own-rooted vines that might be susceptible to phylloxera. It was all about cost. The pioneers had to make many difficult decisions like that.

We did learn some lessons from other agricultural players in the Willamette Valley. There was a robust orchard industry in the area consisting of cherries and hazelnuts. So there were local resources of information about how to farm these crops. One of the practices of the cherry guys was to "clean cultivate," meaning complete removal of grasses and weeds under the trees. It was much the same with the hazelnuts. Farmers completely cleared the ground under the trees and waited for the nuts to fall. Then they used vacuums to harvest them.

So that is what we did, at least at first. Having studied soil science in college, I was aware of the long-term impacts of soil erosion and the benefits of having a cover crop of plants that hold the soil together. It was not until a heavy rain in 1982 when I found about five inches of mud over the road at the bottom of a hill that I realized it in living color. We needed a cover on the soil. But how much cover and when it was needed and how to plant it and care for it would take some time to figure out.

My mathematical skills, derived from a father who was an accountant and a mother who taught math, would prove useful in many aspects of grape growing.

One was crop estimation, which is important for several reasons. The Californians did not really try to estimate the size of their crops before harvest, partly because they had a longer growing season. They could just wait to see how things turned out and scramble to find the labor and equipment they needed to harvest and process the grapes. And because they did not emphasize quality as much as we did, if a California grower had an extra five tons of grapes, he could just sell them to Gallo or another high volume producer.

We had to get smarter. Our weather turns cold and rainy in October, and we cannot get our equipment into all the muddy rows on all the hills. If we don't get there in time, the grapes become diseased and start to rot.

Disease control in Oregon was, in general, more problematic than in California due to our weather. We experienced some years when the clusters developed botrytis, or grey mold, because it rained during harvest. We could see that when the clusters were covered by leaves, they were more prone to get the mold. So we started pulling off some of the leaves around the clusters to better ventilate them. Too many leaves prevented fresh air from reaching them and increased the risks of mold. We did that by hand. (Machines are available to do the same thing today.)

With a vague sense of Burgundian custom, we felt that the highest quality would come from low-yielding vines. And we knew that we needed high quality. We knew that our yields would be lower than warmer regions, and therefore our costs would be higher. If our costs were higher, our bottle prices would have to be higher. If our bottle prices would be higher, it had to be high quality. So there was a sense that anything over three tons per acre could reduce quality. Later, weather trends and further research would challenge that assumption.

One of our goals was making sure we had the right amount of equipment for the grapes we were going to harvest. The ideal was having all of the grapes harvested and all of the fermenters and barrels full. If they were full and all of the vineyard was not harvested, then the remaining grapes would have to wait in the field while winery workers processed the grapes harvested earlier to make room for more. That meant the grapes in the field might have to sit through a few rains before being harvested, and that would impact the quality. Time of harvest is one of the biggest factors in a wine's quality. Sometimes logistics determined that, not artistic or scientific judgment. The flip side of the equation was that we did not want to harvest all the grapes and find that we had extra barrels or fermenters that sat empty. That drove the accountants crazy. We were wasting money.

Lastly, we were to learn that the quantity of wine harvested had to fit into a sales channel. If we tried to squeeze too much through that channel, we might be forced to discount the price and that would hurt brand image.

What we did not know was that we were just riding the coattails of Mother Nature—and she did not care about our precious wine. The grapes existed for the purpose of reproducing and making baby grapes. There is an all-important event every June when the grapes bloom. During the bloom, a percentage of flowers are converted into an immature berry. That percentage depends on the internal health of the vine but also the environmental factors occurring during the bloom event, namely weather. We sometimes have rain at that time of year.

When conditions are great and a high percentage turn into berries, a cluster of grapes will weigh more than when the weather is bad and a low percentage turns into berries. Years later, we would document that the range of cluster weight from one year to the next was from 40 grams per cluster to 140 grams per cluster. In the 1980s, we had no way to determine in advance where in that range we fell in any given year.

The year before I started at Knudsen Erath in 1979, for example, the Knudsen vineyard produced sixty tons of grapes off of sixty acres. Some of that had to do with deer grazing in the top half of the vineyard. The following year (my first year in 1980), we produced 125 tons, which was still sub commercial, meaning not commercially viable. After receiving some pressure to improve yields, I overshot it in 1982 with three hundred tons. There are lots of logistics involved in producing swings like that. You needed people, trucks, bins, tanks, barrels, and forklifts. If you overloaded the system with too many grapes, it would break down.

Lastly, all the things we did not know combined with some up-and-down weather to create an inconsistency of quality during the early '80s. That was tough at a time when we were trying to establish ourselves on the world stage. In 1982 and 1984, we could barely get grapes with over 18 percent sugar by the time the fall rains arrived, which was subpar. (There's a rough correlation between the sugar percentage in grape juice as tested in the laboratory and ripeness. The more sugar, the more ripeness, the more flavor. The ideal range is 22 percent to 23 percent sugar. (The sugar percentage is called the Brix.) Several producers had to make a Rose out of their Pinot grapes and sell the wine for a much lower price. It was not what they wanted to do. It was hard to promote oneself as a top tier Pinot producer while pushing out a Rose at $4.99 per bottle.

But in between those two years was 1983, which I think was Oregon's best vintage ever. By that year, enough vineyards were coming into production that there was enough wine to go around. It was that vintage that Ken and I tasted out of the barrel when we snuck into Erath's cellar during one of Ken's visits to Oregon in 1984.

We looked at each other and said, almost in unison, "That's it."

Ken Wright adds:

"Back at the time, in the '70s and early '80s, when I was in California, and Allen was up in Oregon, we would travel to meet with each other, going in both directions. So I had the opportunity to taste Oregon's Pinot Noir from the bottle as well as from the barrel.

"At that time, most people outside of Oregon shared the opinion that Oregon wine was only as good as Mother Nature was kind. Their meaning was this: in the years that were warmer and drier, and when the rains held off until the later part of October, the wines were compelling and beautiful and detailed and had a lot of intensity. But in years that were cooler or when there was ill-timed rainfall, the wines were just okay. The industry was living and dying on the basis of weather patterns.

"But during those years, I had some Oregon Pinots that were glorious. It was like getting a bit of religion. What I loved about them was the profile, which means both the taste and the aroma. The profile was exactly what I was hoping for in a Pinot Noir. Rather than being austere or being overwrought, which used to happen in California, the fruit quality of Oregon Pinot Noir in the good years was absolutely on the mark.

"It was like going to a farm stand in late summer when the fruit reaches its epitome of ripeness. The Oregon Pinot grapes were so beautifully spot-on. You hit ripeness head on. Not overripe, not underripe, just stunningly perfect.

"They made me a believer because we knew what was possible even though at the time it was not consistent. It clearly could be done. I was impressed enough that I decided to move there in 1986 and to focus on Pinot Noir, period. To me, it's absolutely world class. When you tasted the wines, you understood this was where place and plants were perfectly matched.

"It's like eating San Marzano tomatoes in Italy. They are, without question, the best in the world. It's because that plant loves that place. Those tomatoes want to be great. Or eating bamboo shoots in Kyoto, Japan, during the cherry blossom festival. I felt the same way about Oregon's Pinot Noir grapes."

Rollin Soles took a different, longer path to the Willamette Valley. Originally from the Dallas-Fort Worth area, he was first exposed to vineyards and wineries when his family relocated to Spain when he was in elementary school. Then later, when he was a student at Texas A&M, he spent ten weeks one summer working at Schlossgut Bachtobel, an estate in eastern Switzerland in the shadow of the Alps, according to a profile in Texas Monthly. He learned about what he called "farming on the edge" because of the difficulty of getting grapes to ripen in such a cool climate. Ironically, that helped prepare him for Oregon. "Whenever things get really tough (in Oregon), I just think about Switzerland," he said.

After finishing undergraduate work at Texas A&M, Soles attended UC Davis to get a degree in enology, the study of wines. He made his first trip to the Willamette Valley while a grad student in 1979. "I drove into the Willamette Valley, and I immediately knew this was the place I was meant to be," he said.

But it did not happen overnight. He worked in Napa, once more in Switzerland, France, and eventually Australia. He followed a girlfriend to Australia where he met Brian Croser at his winery, called Petaluma Winery. Croser, also a UC Davis grad, offered him the position of chief winemaker. But it didn't last forever. "The geography and the early wine-makers convinced me to return and build a career in the Willamette Valley," Soles says today. His relocating to the Dundee area would prove to have enormous consequences.

CHAPTER THREE

The Momentum Starts to Build,

1985–1990

WHEN KEN WRIGHT HIT A cross road with his situation in California, he decided to turn right and move his family to Oregon in 1986. He had had his eye on Oregon ever since we tasted the '83s and knew that is where he wanted to make his mark. He had made some wine off the books the year before and now wanted to sell it to finance the move. The problem was that one had to have a license to make that much wine—and he did not. Somehow, using his amazing persuasive abilities, he convinced the federal authorities to allow it. He kept talking his way out of problems. I was not surprised because I had once seen him talk the repossession man out of taking his car away.

But I worried about him. It was a big leap. He had kids. We drove down North Valley Road looking for a barn for him to rent. We were not smart enough to get a realtor. Nor did we know if Oregon's land use laws would not permit it. We were just going to go knock on doors like good ol' Kentucky boys. Ken eventually rented a place in McMinnville, a nearby town, where he started Panther Creek Winery.

Rollin Soles had known Ken Wright at Davis and moved to Newberg, just a couple of miles from Dundee, in 1985. Having worked in so many places, he brought real experience to the game. Now in Oregon, he was prospecting for land and trying to help Croser decide whether to take the plunge. Soles began to talk to me about getting involved with them in a sparkling wine operation. Rollin and I had science in common.

People kept coming. A friend called me and told me he had met a couple from St. Louis who were interested in the Oregon wine business. Did I know a place they could rent? I did. My neighbor Gary Fuqua had a basement apartment at his vineyard home. The couple turned out to be Bill and Deb Hatcher. He had had a career in corporate finance and was tired of the treadmill. We became neighbors and would walk through the vineyards to share dinner at each other's homes. After they had children, Deb nicknamed their basement apartment "diaper dungeon." Bill would later become the general manager at Domaine Drouhin and eventually co-owner of A to Z Wine Works.

Don and Wendy Lange were inspired by tasting an early bottling of Knudsen Erath Pinot Noir and sold their suburban Southern California home to buy twenty acres down the road from me. Don had been a folk singer, and the couple had no previous experience in the wine business. They had a slow start before planting their own vineyard. But they helped raise the bar in hospitality by raising the professionalism of their tasting room and proving that direct-to-consumer sales could work. Dundee was gaining momentum as were several other towns.

Croser was a very charismatic guy in addition to being technically driven. He came from Australia to visit during the harvest of 1985. It was a dry year, and some of the leaves were senescing, or gradually deteriorating biologically. He said, "You need irrigation." He pointed out the yellow leaves that were caused by lack of water. If we wanted to improve consistency, he said, we had to consider all aspects of vine balance. That led to a whole experiment in subsequent years with irrigation that remains somewhat controversial in the industry because of water use issues. But Croser at that time was the tip of the spear of people from other parts of the world with more experience in wines than we had in Oregon.

Another important personality who spent time in Oregon in 1985 was Véronique Drouhin, the only daughter of Robert Drouhin, the legendary Burgundy winemaker. She was both fun-loving and hard-working. She worked as an intern at Adelsheim Winery, where I met her.

Vero, as we called her, was the third generation of her family to work in the wine industry. Here is the start of her story in her own words:

"MY FATHER LOVED THE AMERICAN people he worked with. My father has always loved the US. He has been going there from a very young age in the 1950s. The relationship of the family and the United States goes way back. My grandfather, Maurice, was general liaison with Gen. Douglas MacArthur, then still a major who conceived of the idea of creating the Rainbow Division to get American troops to Europe in World War I. Maurice spoke good English. He was very involved in the army. But he also had a little company making wine.

When I studied oenology at the University of Dijon back in 1984, I was the only girl in the class, and that has never been a problem for me. During our studies, we were required to do an internship. I did one in Bordeaux because I felt that was a region of France I should know better. I thought California also would be fun. (I didn't think about Oregon.) My father was a very good friend of Robert Mondavi, the California winemaker who came to Burgundy on a regular basis. I had known him since I was a child. He has done a lot for Napa Valley. And he was the nicest man.

"This was 1986. My dad said: 'Of course, I think Robert would take you as an intern. But if I were you, I would go to Oregon.'

"'Where is it?' I asked. 'Why would I go there?' I didn't know where it was. I had no idea they made wine.

"'Well, I have been to Oregon to sell my wines from Burgundy because we have some distribution there and good customers,' he said. 'While visiting, I have tasted some very interesting Pinot Noirs." He did not say all of them were good, but he said that most of them were potentially very good. There may have been some technical issues. 'I think Oregon is a good place for Pinot Noir,' he said. 'Just go and have fun.'

"We asked our distribution company to find wineries who would accept a young girl from Burgundy who had very little experience and didn't speak much English. Three of them kindly accepted me for the 1986 vintage: Adelsheim, Eyrie, and Bethel Heights. I could help all of them make wine.

"Before I started the internship in the fall, we went for the summer of 1986 with my parents and with Laurence Jobard, one of the first woman winemakers in Burgundy who worked with us from 1974 until she retired in 2006. The world of Burgundy wine was very masculine. But she inspired me when she said, 'If you like wine, pursue your winemaking studies.'

"I spent four incredible months in Oregon in the fall. What I saw was this: It was an extraordinary experience. I was very impressed by the people who were so courageous to start making wine with so little equipment. After this time there, I came back for Christmas to Burgundy. Shortly after, David Adelsheim called my father to see if he would be interested in making wine in Oregon.

"I remember my father saying, 'It would be fascinating. I believe in Oregon. There's great potential, but I don't have the time.' For our own family company, he was the winemaker, he was the vineyard manager, he was managing the company and he was traveling to sell the wine. So he had a lot of jobs. My three brothers and I, as young adults, were just starting to join the business."

Dick Erath and Cal Knudsen had been partners in one of Oregon's largest wineries for almost fifteen years when differences began to appear. Erath was the onsite manager, and Knudsen was the off-site investor. They had different visions. During the time in 1986 when I was negotiating with Argyle and building a house, Knudsen informed me they had decided to terminate their partnership, and he was not sure if he was going to buy out Erath or the other way around. It was a time of uncertainty for everyone.

Knudsen and Erath had devised a very civilized method for dissolving their partnership in the winery. Party A could end the partnership by offering to pay a price to Party B for half the business. Party B then had the right to sell his 50 percent of the business for that price or buy the other partner out at that price. Up to this point, all of the grapes from Knudsen's sixty-six acres of vineyard went to Knudsen Erath. If that partnership went away, it meant that some grapes could be sold elsewhere. For many of the people interested in investing in Oregon, one of the first problems was sourcing high-quality grapes.

In the end, Erath decided to buy Knudsen out. Knudsen owned the land and the buildings, so they arrived at a carve-out lease deal that left Erath Winery operating as an island in the middle of the Knudsen Vineyard land. Part of their separation agreement was that Knudsen would continue to sell grapes for a period of time to Erath.

At the same time that the Knudsen Erath partnership was dissolving, I was talking to Argyle. They wanted to offer me an employment contract. That sounded great after what I had been through at Knudsen Erath. Brian Croser and a partner from New York were in town in late 1986 and delivered a employment contract. They left town, and I read it. It had a noncompete clause in it that caused me a lot of heartburn. I felt like I had been in Oregon before them and would be here after them. I called Rollin Soles, who was going to be the winemaker at Argyle, and told him the deal was off. I was upset.

The next day, I was at the cabin and Soles and Croser walked up on the deck. Croser had flown back from San Francisco to talk me down. He said, ignore the non-compete clause—it was just some asshole New York lawyer playing games. I never signed the contract but agreed to start work without one. My relationship with Argyle came that close to not happening. (It lasted for more than thirty years.)

Then it was up to me to talk to Cal Knudsen. I was shaking when I called him in Seattle because I might have had to tell him our relationship was over. On the phone at first, I said only that I wanted to come to Seattle to talk to him.

"Why, what's up?" he asked.

He got it out of me on the phone. I told him I wanted to join a group making sparkling wine, and they were interested in meeting him. "I'd love to meet them," he roared. Rather than terminating our relationship, we were expanding it.

That brief conversation would lead to him becoming an investor in and supplier to Argyle (when his supplier arrangement with Erath expired). We also talked about my continuing as his vineyard manager because Argyle wanted to start a vineyard management company to assist in sourcing grapes and subsidizing their launch. We had lunch at the Arlington Club in Portland, where he was a member, to discuss it. He agreed to keep me on by becoming a management client of Argyle, where I would be employed.

From left, Allen Holstein, Cal Knudsen, Ian McNee and Brian Croser at the first planting at Lone Star Vineyard in 1997.

HOW I ENDED UP ON MY HILL

I WAS JUST A HIRED hand with no tangible stake in the wine industry other than my measly wages and discounts on wine. It was Cal Knudsen who explained to me one day that I needed to own something. We were out walking in his vineyard, and out of the blue he said to me, "You know, Allen, for this to work out for you, you are going to have to own something, and I will help if you can find something." That conversation led me to believe I could, in fact, own something.

It happened because I was moonlighting for growers such as Gary Fuqua, helping them out on different projects. As a result, I heard that he was getting divorced and he would have to sell part or all of his vineyard. There was no shortage of divorces and remarriages taking place. The insider's joke was that a little ego, a little wine, and a little sex, combined with a lot of money, made people behave differently than if they lived in a three-bedroom home in the suburbs and commuted to work every day.

Gary Fuqua and I agreed on a deal where he would finance the sale of ten acres at the top of his property in the Dundee Hills to me, and he would keep his house and bottom of the vineyard. If he defaulted to his bank, they would foreclose on my section as well. But there was no way I could get financing in my name. My section was planted and producing grapes, so I would have income the first year. I threw in every nickel and bought it in 1984. I was no longer a hired hand. I owned something.

As the partnership began dissolving at Knudsen Erath, and as I continued talking with the soon to be Argyle group, I realized that there was a scenario where I might not work for Cal Knudsen and therefore could not live at the cabin (because it was on his land). So I decided to build a house on my vineyard property and moved in in December 1986. I lived there until 2021, when it became the tasting room for Holstein Vineyards and Granville Wines, the label launched by my son, Jackson, and his wife, Ayla. Nearly every visitor is impressed by the sweeping, panoramic views of the vine-covered hills below, all overlooking the Willamette Valley.

Argyle focused initially on sparkling wine because they believed they would achieve better consistency than the pioneers had with Pinot Noir. The grapes for sparkling wine did not have to be as ripe. Sparkling wine lifted the game for Argyle's red wine because it allowed them to only use the ripest grapes for the red. Other wineries followed suit when mobile bottling units began to appear that had specialized equipment for bottling sparkling wine. In fact, we did not make Pinot Noir until techniques improved.

Argyle vastly overestimated demand for sparkling wine. They made twenty thousand cases a year for three years before releasing any for sale and essentially had to give it away. It was a new winery from a new region with a new product. But at the time, they did not know that. Their concern about getting started in Oregon was a source of grapes. They felt the viticulture was too immature to buy land and use current techniques. So their strategy was to lease other vineyards and seek to build up inventory quickly, making it a more efficient use of capital. With matters at Cal Knudsen's vineyard up in the air, Brian Croser wanted me to find existing vineyards he could lease. So I was happy to lease them my vineyard and introduced them to other growers I knew. All of these guys had supplied to Knudsen-Erath, and none were too happy with the prices Erath was offering for their grapes. I felt like it was a good deal for these guys, most of whom I had helped in their vineyards. They agreed to lease some or all of their vineyards to Argyle as of January 1987 as I was starting at the company.

Allen Holstein and Cal Knudsen walking the site of Cal's beloved Pinot Meunier.

Robert Drouhin, Bill Hatcher, Veronique and me in the late 80's in my dining room discussing the future.

GLOSSARY

ROOTSTOCKS — Rootstocks do not have any direct impact on the quality of the grapes. It's more subtle. Some rootstocks are de-vigorating, meaning they don't support robust vines. Other rootstocks create excessive vegetative vigor, which was one of the early problems in Oregon. That meant the plants put too much energy into growing more vines and not enough into ripening the grapes. The goal, of course, is to use rootstocks that create ideal growth and that can tolerate insects.

CLONES — Any clone can be grafted onto any rootstock. The type of clone used has more of an impact on the taste of the grapes than does the rootstock. Clones are to a winemaker what the colors on a palette are for an artist. They give the experienced winemaker blending opportunities.

VINE — This term refers to both the roots and the grape variety it may be grafted to. It is the total plant.

GRAFTING —This is done with a machine that mounts on a table and makes reciprocal cuts in the rootstocks and the clone variety so that they fit together like a bone. The vine develops calluses and then heals. It is reminiscent of how a surgeon repairs a human bone.

VITIS — This term refers to the genus, or broader family, of grapes. All grapes are Vitis. Some are native to Europe, others to the United States. Vitis vinifera is a member of the Vitis genus.

AXR — This was a rootstock that was a cross between American rootstocks and European species and used widely in California. It produced higher yields, but it had vinifera in its DNA, which ultimately rendered it susceptible to phylloxera, a soil borne insect native to North America.

SELF-ROOTED — This term refers to a situation in which a vine and its roots occur naturally together. It is not grafted. In the early days in Oregon, the pioneers planted all self-rooted grapes. But over the years, we experimented with different combinations of rootstocks and clones to maximize flavor and reduce disease. Today, very few of the grapes used to make wine are self-rooted.

Brian Croser hosted me in Australia in March of 1987, so I could look over their harvest procedures at Petaluma with an eye toward adopting them in Oregon. I got to see Hunter Valley, Clare, Barossa, Coonawara, and Picadilly. I met a lot of his crew who would come at one time or another to Oregon to assist with harvest. Piccadilly was the coolest region in Australia. We argued if Oregon or Picadilly was cooler (and by extension more challenging). I could see the level of investment in Picadilly far exceeded anything in Oregon. And because the Australians had continued their relationship with the British and by extension the Europeans longer than the Americans did, there was a sophistication in the trade that was absent on the American West Coast at that time. The big French champagne brand Bollinger was a partner in Petaluma, for example. Croser had spent considerable time in the Champagne region of France and brought some of that knowledge with him to Oregon.

In Dundee, Rollin Soles found a warehouse behind an Afghan restaurant and city hall and adjacent to a warehouse supplying mobile home construction materials. His job was to prepare the building to become a winery by pouring new concrete floors with proper drainage, ordering the equipment and supplies in readiness for September harvest. Over the course of the year, progress was slow, and 1987 was the warmest year of the decade, so it would be an early harvest. I was worried. Only later did I learn that at least one of the business partners was slow in coming up with the promised cash to get the place ready in time. That had been the hold-up.

At the last minute, things began to happen in August, and the concrete was finally poured. We started harvest on Labor Day weekend, the earliest ever. We had to pay air freight to fly in the wine press from Germany. We started harvesting grapes with the press still literally in the air. But Soles pulled it all off. "Having been one of the first 'flying' winemakers in Australia, I had some experience in the craft of pulling it all together to meet the harvest," he recalls. "Plus, I had the fantastic help from Australia of winemaker Martin Shaw and winery engineer Trevor Underwood."

Meanwhile, at the same time, Véronique Drouhin had helped pave the way for her father, Robert, to purchase land in the Dundee Hills. His Burgundian company had such an international reputation, and the announcement translated to an enormous endorsement of Oregon's potential. He obviously had great experience in all aspects of the business in France and would bring all of that to bear in Oregon. His purchase increased the comfort level of the other more risk-averse set of investors to follow.

Here's how it happened, in Véronique's words:

"IN JUNE 1987, DAVID ADELSHEIM called my father again, and said, 'Robert, there is a beautiful piece of property for sale in the Dundee Hills close to Eyrie vineyards.' We liked those wines because they were very elegant. Plus, 1987 was the first year that Oregon organized the International Pinot Noir Celebration . . . They invited people from all over the world, including Burgundy. My father was a guest speaker. He said, 'Let's go.' We arrived in July.

"Everyone knew that an investment by Robert Drouhin in Oregon would be big news. They told him: If someone quite famous invests in Oregon, it's going to be important for the whole valley. David Adelsheim and Dick Erath told him that.

"David walked us out the little road to the piece of property that was atop a hill. There were big fir trees that are still there. It was a wheat farm . . . The wheat was blowing in the wind. We were stunned with the view. It wasn't just the view that made us decide to buy the place, but that's when my father said to me, 'If you are willing to take care of this, if you are willing to make the wine and spend a lot of time with this adventure, I think we should try.'

"That week, we bought the land. It was not expensive, something like $3,500 an acre. 'Okay,' he said, 'We have the land. We should make wine.'

"I said, 'Alright. But we have no winery. We have no equipment. We have no grapes.'

"'Okay,' he said, 'We will buy grapes, bring equipment from Burgundy and try to find a place to make wine.' We ended up renting a newly built small warehouse at Veritas winery. It had no water and no electricity. We got some electricity plugged in, and a hose was available. I had to find the rest of the equipment. The next big thing I had to do was to buy a press for the grapes. I went to California. I was only twenty-five years of age and did not have much experience. But I bought a press! Robert's idea at first was to buy grapes from different locations. We didn't know which one would be best. We bought from Knudsen, Durant, Hyland, Canary Hill, Seven Springs, Bethel Heights, and

Forest Grove . . . We had an interesting variety of sources. This was a great way to start making wine quickly.

"We had no employees except Bill Hatcher, who had been hired to take care of everything. For thirteen years, he was the manager of Domaine Drouhin Oregon (DDO). He helped a couple of the first harvests. Otherwise, it was just my parents, myself, two cousins from Belgium, and Jean-Yves, an engineer from Burgundy. That was the team in 1988.

"Father knew he wouldn't make money for a long time—and he was right."

Drouhin announced his purchase of more than a hundred acres in August 1987 as we were ramping up for harvest at Argyle. The sheer size of his acquisition was impressive. The rumor was that he would plant vines in 1988. Having begun a vineyard management business under the umbrella of Argyle, I was disappointed to not get an audience with him.

Then one day in February 1988, after the harvest was long over, Bill Hatcher, my former neighbor who was now the general manager at Domaine Drouhin, called me early in the morning and asked if I would like the vineyard manager's job at Drouhin. I could either do it as an employee of Drouhin's or as part of Argyle's vineyard management business. I told him Brian Croser was in town, and I would ask that day. He asked, what if Croser said "no." I said I would take it anyway as a Drouhin employee. But I met with Croser, and he was all for it. I called Bill Hatcher back and accepted. Oregon's first vineyard management company was being hatched.

Drouhin had organized to plant own-rooted vines in the spring of 1988, but Robert Drouhin was in Burgundy. Where on the property did he want to plant and at what spacing? We were going to plant a first-of-a-kind five-by-seven-foot spacing at Knudsen's with drip irrigation that year. That was 1,200 vines per acre. Did he want to do that?

I got a handwritten note sent via fax machine explaining that in Burgundy they were pretty certain that a meter (slightly more than three feet) between vines was important. He had consulted Burgundian colleagues. But they could not decide on the distance between rows. He would let me decide that based on logistics. So I used his one meter between vines and I chose seven feet between rows because I could get a tractor narrow enough to go down seven-foot-wide rows. Before 1988, one could not find a tractor that narrow. John Deere introduced a vineyard model that year made in Germany. So in a matter of weeks, Knudsen completed a five-by-seven-foot planting and Drouhin did a one meter by seven foot planting, both firsts on the West Coast. We were experimenting. The five by seven became the industry standard in premium regions of the West Coast.

Cal Knudsen was not a technical guy, but he could tell that Drouhin and Croser were really smart about the science of grapes. Not only did he want to support them by selling grapes to them, he had another agenda, which he explained to me one day, "You watch what they do in their vineyards, and then do the same in mine!"

The arrival of bigger, smarter money was a pivotal moment in the Oregon wine industry. In terms of prestige, the Drouhin investment had higher impact than the creation of Argyle. But both investments were in the Dundee area. I participated in this fascinating moment in the Oregon wine industry as a relatively minor player, but my relationships with different people helped it happen. And in this business, relationships are critically important. The foundation of the Oregon wine industry's credibility was established during this era because the Drouhin and Argyle investments fundamentally reduced the risk for others to follow. And I was rewarded personally as well, obtaining my own vineyard and leaving the cabin on Knudsen's hill for a real house located on my own vineyard.

Another major development in the Oregon wine industry that started at about this time was the move away from buying grapes by the ton, adds Ken Wright. If a winemaker is buying grapes by the ton, the grower has an obvious economic incentive to produce as much tonnage as possible without paying a great deal of attention to the quality of the grapes. "For Oregon to succeed, we realized we needed a different way of doing business," he says. "Let's look at the vine like an automobile. Your leaf surface is the photosynthetic engine. You're asking that engine every year to ripen a crop. In a cooler year, you're going to struggle if there is too much crop because the engine is being asked to produce more sugars for all that weight of clusters than the environment will allow. We knew we had to change the paradigm.

"Allen gets into this subject in the next chapter, but we knew that if we dropped crop, meaning cutting off clusters of grapes, maturity would happen faster. You need to act decisively and remove the weight from the plant to get to the correct level where the plant will be able to ripen that crop fully before you get to mid-October.

"The new way was to buy grapes by the acre. We went to the growers and said the current way of doing business is a broken dynamic. We need to be successful in all years, not just some. The world doesn't accept 'just okay' wine. The wines coming from people who did business in this way, even in the cooler years, were noticed by a lot of people. These wines were compelling and intense and nuanced and beautiful.

"But it doubled my fruit costs. We had to raise prices accordingly. We had to take a risk that people would pay for it. We believed that for the quality we were going to have, and the consistency of quality, that it would be worth it. People would pay for it. And they did. And they still do."

Domaine Drouhin made quite a statement for brand Oregon when it was built in 1989

Allen Holstein and Rollin Soles in our prime at the International Pinot Noir Celebration

THE FRENCH CONNECTION

THE DROUHINS INVITED ME TO Burgundy in January 1989. I thought it was going to be ten leisurely days of two-hour lunches and lovely scenery—but it was a real work trip. Robert had decided to plant one of the first European-style high density vineyards in the New World that year. He knew he would have to get a French straddle tractor to do it. These tractors are tall enough to travel over the top of the vines because the rows are too narrow for a serious piece of machinery to get through.

Robert drove me out to one of his blocks of vines where two guys were working on a tractor. He asked in French if the American could drive it through a block of premier grapes. I jumped on, and the two guys jumped on, smoking French cigarettes and yelling instructions over the sound of the engine. I was half-terrified but made it out and back.

Satisfied, Robert drove with me straight to the factory in Beaune that makes the tractors and sat down with some engineer types while they droned on about the design of the tractor they would make for Oregon in French. When they had a question, they would switch to English and ask me a specific question, the genesis of which I knew nothing about. I was faking it.

Then it was off to visit French nurseries to learn how they produced grafted vines. Robert wanted to plant grafted plants in Oregon. He was the first to do that also. He wanted to get to the point of producing his own plants from his own rootstock blocks. There was no expertise in Oregon at that time, and he wanted me to learn it. We saw rootstock blocks out in the French countryside, and he went and knocked on the door to introduce himself and ask if we could see how they did it. They recognized his name and were most helpful.

There was one embarrassing incident when I turned right on red in front of a cop in one of Drouhin's vehicles, not a violation of Oregon law but clearly a violation of French law. I got hauled into the police station for a good dose of harassment before they called Robert to see if I had stolen the car.

Robert could tell I was overwhelmed with it all. When it came time to leave, he drove me to the train station, bought me a ticket to Paris, and walked onto the train and said 'sit there' along with instructions of what to do when I got to Paris. I was never so happy to see an American Airlines jet sitting at the airport waiting for me. I didn't realize it at the time, but I was acting as a kind of funnel for French wine expertise to reach Oregon.

Shortly after my return in January 1989, construction crews showed up to begin construction on the Domaine Drouhin Oregon (DDO) winery, which would be located on the vineyard. The idea was to have it ready in time for the 1989 harvest. Drouhin had planted 1 meter x 7 feet in 1988, but in the next year his conviction grew to make the leap to true high density of 1 meter x 1.3 meter (roughly four feet). That would require three thousand vines per acre compared with five hundred in the older plantings at Knudsen.

He asked in advance, could I plant thirty thousand vines on ten acres? At that time, it was difficult to get labor once the strawberries ripened in early June because all the Mexican laborers got busy with the strawberries. I told him, "I could build the pyramids in May, but June is another matter." He chuckled.

We planted Drouhin's high density vineyard in May 1989 with grafted vines. This was a first for Oregon and the New World. There was some competition between him and another major European family, the Rothchilds, who owned Opus in Napa Valley, but Drouhin was first. His French straddle tractor was delivered in June along with a Frenchman who spoke no English but whose job it was to explain to us how to operate it. There were no others in the entire U.S. We had fun with it. I took the Frenchman to a Western rodeo, and he cracked up. He had never seen anything like it.

Drouhin was not sure at first about the high-density Burgundian-style plantings in the New World. One day in an effort to reassure himself, he mused out loud to me, "It will be better viticulture. We will either get better yields or better quality or maybe both." In the end, I came to agree that he was right—we got both.

Shortly after the harvest of 1989, Brian Croser announced to Rollin Soles and me that he was going to buy out his partner in Argyle. That proved to be the opening for Cal Knudsen to get more involved. Argyle became a partnership of partnerships. Acting in his personal capacity, not as part of Petaluma, Croser brought in a silent partner, and Cal Knudsen brought in some well-heeled Seattle investors in the form of a partnership called the Dundee Wine Company.

Cal Knudsen would be chairman of the Argyle board. Rollin Soles and I were given seats on the board even though we had no financial interest. It felt like we had a say in the business.

As an industry, it had basically taken us a decade to reach a point that our success appeared to be possible. The decisions by the Drouhin family and the Argyle group to establish vineyards and wineries captured the attention of many others who assumed that such established and experienced groups would succeed. Not coincidentally, land around Drouhin's Dundee vineyard began to be sold, and before long, Archery Summit appeared next door. Domaine Serene bought the next hill over. The Drouhin purchase increased the confidence level of would-be participants. All brought experience and ideas that lifted the game for Brand Oregon.

The Knudsen-Argyle-Drouhin alignment was a win-win for everyone involved. Before it was over, Cal's vineyards were supplying grapes to both Argyle and Domaine Drouhin, who was buying his grapes because of the sheer amount of wine they intended to produce and also because

Drouhin knew it was going to take time for his grapes to get established and start producing in quantity. Douhin had an import company with relationships with distributors nationwide. They were going to handle sales of Drouhin's Oregon wine, and they agreed to handle Argyle as well. So there was a connection at the marketing level as well as at the viticultural level.

At first, the Drouhins' friends back in Burgundy thought they were crazy for making wine in a place called Oregon.

In Véronique's words:

"IN 1988, WE MADE A hundred barrels of Pinot Noir. We didn't release it commercially until 1991. We asked ourselves, 'Did we like the wine?' We did. Then we asked ourselves, 'Are we the only ones who like it?'

"We brought samples back to Burgundy and shared it with friends and competitors and very fine winemakers with great palates. We asked them to not be polite and to be very honest. The response was truly encouraging. But on the other hand, they said, 'You are totally crazy. Nobody has ever heard of Oregon. Good luck.' The good news for us was that the quality was there.

"In 1989, my father said, 'We should have our own winery.' It was a very exciting moment to think of how to design a 'dream winery.' He and I spent a lot of time thinking, where did we want to build the winery on the property? We wanted to use the natural slope of the hill to build a gravity flow winery. It ended up being on the upper part of the hill where we had both the stunning view of the valley, and we could use the natural slope.

"I really loved the Oregon system because you are not tied by regulation. If you want to blend this fruit with that fruit, it's fine. I liked the freedom we had and still have. There is freedom to make the best possible wine. In Burgundy, everything is regulated. It makes things a bit more complicated.

"It takes time to see if the wine you make is consistently very good from that location. Is it due to the soil, is it due to the vintage, is it due to the clone, is it due to the rootstock? Is it a combination? It doesn't happen after one, two, or five, or ten years."

In my own personal life, It felt like I had achieved a level of stability that I had not enjoyed in almost nine years since arriving in Oregon. A new chapter of professionalism was upon me. I was managing about two hundred acres of grapes—my own vineyard, the Knudsen vineyard, Drouhin's vineyard, and Argyle's leased vineyards. The Dundee Hills were emerging as the epicenter of the Oregon wine industry with all that implied for real estate values.

It's worth noting that Brian Croser introduced a new technique from Australia that improved the early Oregon Chardonnays. In making white wines, oxidation of the grapes before they are crushed can be a problem. It's like if you cut an apple and leave it out in the kitchen. It turns brown, which impacts its flavor. But if it's in a refrigerator, it stays fresher longer. On that principle, Croser invested in giant coolers at Argyle to chill the grapes before they were pressed. He also chilled the juice after the grapes were crushed, thereby preventing fermentation from occurring until after the harvest was completed. That allowed the staff to focus on one fermenter at a time, improving quality control. As a result, Argyle's

Chardonnay from 1987 was named one of the Wine Spectator's top hundred wines of the world at a time when Oregon had a pretty rotten reputation for Chardonnay. It was a breakthrough. As more and more wines from Argyle, Drouhin, and other producers started hitting the market in 1990 and 1991, Oregon started to find traction, at least regionally.

Argyle did not actually produce a Pinot Noir until 1992 because of its concern about consistency. "It was a combination of farming for consistency and capacity," Soles explains. "But the market was growing for Pinot Noir wines. Once we were satisfied that a consistently high quality Pinot Noir could be made, Argyle added this important wine to its range." The first "Reserve" level Pinot Noir was from 1993, sourced from Knudsen Vineyard fruit.

Ken Wright was one of the early apostles for Oregon Pinot Noir, and it was tough sledding. "People were unimpressed, and it hurt us," he explains. "Their image of most Pinot Noir was that it was very light and did not have not a lot of intensity in anything—in color, aroma, or flavor. I went to Texas and paid a call on the largest wine retailer in Dallas. I had a variety of beautiful Pinot Noirs to sample with him, but his bias was so strong. He was a nice guy but his mind was made up.

"'You know, Ken, these wines are really pretty,' he drawled. 'But do you have anything other than women's wine? My clientele drinks Cabernet.'

"There was a huge bias in America, especially in Florida and Texas. If you didn't have Cabernet, they wouldn't even talk to you. Now those are my two best markets. That's how much has changed. It's revolutionary. It's the maturation of the American palate."

The Drouhin family also served as important ambassadors for Oregon wine. In Véronique's words:

"FOR US, WE MADE 1988 and 1989 vintages to see if we liked them. And we did like them both. In 1991, my brother Philippe, father Robert, and I went on a tour for two weeks to thirteen American cities. It was a fun road trip to show off the wines. We had three from Burgundy and two Pinot Noirs from DDO (Domaine Drouhin Oregon). People would have a list of the wines, but they did not know the order in which the glasses would be served. In other words, they were tasting blind. The idea was that they would be very honest with their judging. It turned out that people were confused and not able to tell which wine was from where. We told them, 'We are not trying to copy Burgundy. This is Oregon. It's unique. It's different from Burgundy. The soil is different. But it is Pinot Noir, made by the same family that makes wine in Burgundy in a very elegant style. Our motto became, 'French Soul, Oregon Soil.'

"I wouldn't say Oregon wine had a bad reputation. They just did not have any reputation. People just didn't know about Pinot Noir . . . Nobody was thinking of Pinot Noir.

"In 1991, that same year, I was at DDO, and I was cooking. I heard this thing on the television. It was a 60 Minute show called the French Paradox. It asked the question, 'Why do the French, who eat foie gras and all these fatty foods and cream, have fewer heart attacks than we do in America?' They interviewed a professor who said, 'The answer is simple; it's because they drink red wine.'

"I was amazed. This was big. Someone on television in prime time was saying the reason the French had fewer heart attacks was because they drank red wine. It didn't educate people about Pinot Noir, but it instantly got them to be more interested in red wine.

It took about ten years to properly build the distribution network and make people think about Oregon as a serious place. The other big contributor was the movie Sideways in 2004. It was about two guys who got drunk most of the time. They kept saying they were tired of drinking Merlot. The big new thing is Pinot Noir. It had a big impact on the US market."

So a combination of positive media coverage (60 Minutes and *Sideways*) plus tireless travel by all the Oregon winemakers and the reputation of the International Pinot Noir Celebration helped power Oregon's emergence.

MORE THAN 650 WINES RATED

Wine Spectator
www.winespectator.com

Wine Country Travel
Where to Find Fine Food and Lodging to Match Local Wines

- California
- New York
- Oregon
- Texas
- Virginia

+ **Australian Whites: What to Buy**
+ **How to Cook With Wine**
+ **San Francisco Dining**

Rollin and I thought we hit it big to make cover of the Spectator, but the article was about wine tourism, not us!

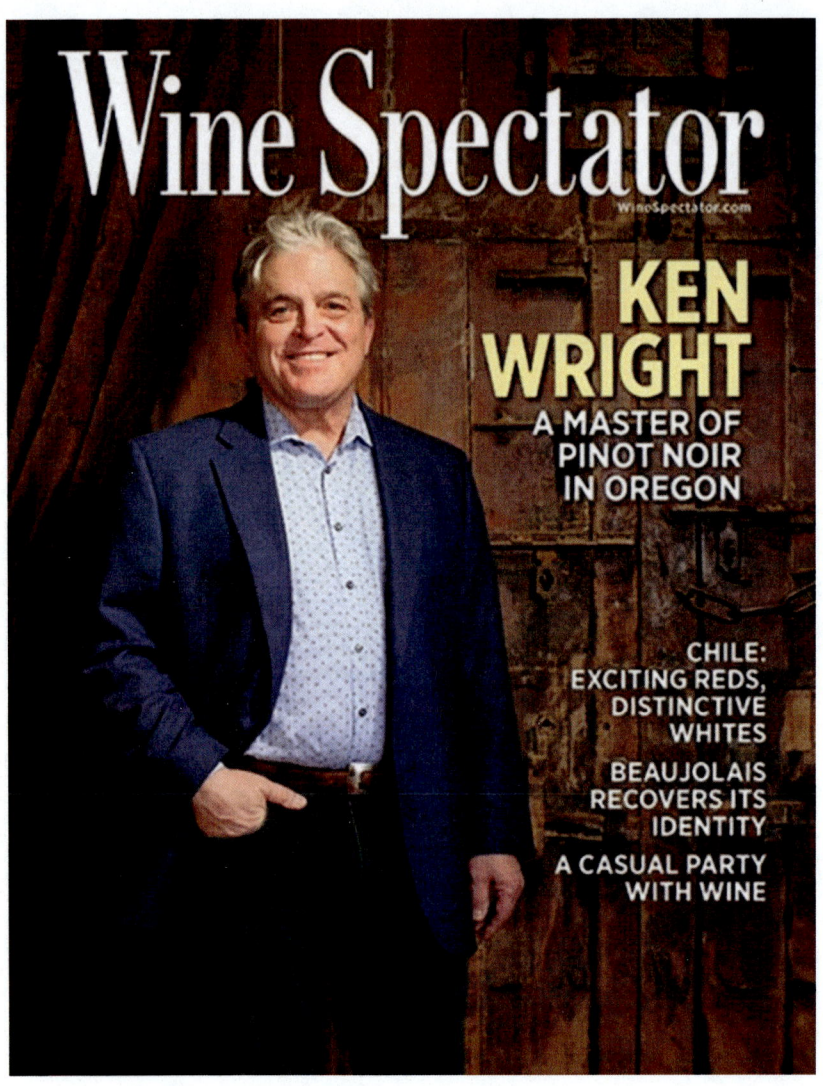

Ken Wright made the cover of the Spectator too but in his case chronicled his achievements.

CHAPTER FOUR

How Phylloxera Almost Wiped Us Out

THE LEGAL PROCESS OF IMPORTING Pinot Noir and Chardonnay clones from Burgundy, France, into the United States was completed in 1989 when the federal government authorized Oregon State University (OSU)—and not the better-known UC Davis in California or Cornell University in upstate New York—to analyze the clones in a sterile, virus-free environment. This was something of a coup for Oregon and stemmed from the lobbying efforts that the early pioneers made to persuade the federal government to tap OSU for the job.

OSU had the job of assuring that no clones were released that were contaminated by any viruses. That was the criteria for release, not the pedigree of the clones as they had performed in Burgundy. They released Pinot Noir clone number 114 and 115 along with a few others that did not have any pedigree. They had Pinot Noir 667 and 777 but would not release them because they said they had some form of virus. The technology was improving, and labs were finding viruses where previously other labs had found none.

Without revealing the details, I came into possession of the unauthorized 667 and 777 and took all four clones (114, 115, 667, 777) to a local mom and pop greenhouse for mist propagation to increase the number of plants. This was a way to get 25 plants turned into 250 plants and then into 2,500—all within a year. I did not know what other material the greenhouse was handling and did not want my material intermingled with any other plants. I labeled them PN 1, 2, 3, and 4 instead of 114, 115, 667, and 777. OSU also released Chardonnay clones 76 and 96 at that time and later clone 95. These clones vastly lifted the game of Oregon Chardonnay.

Many growers got a letter from OSU stating that we could buy twenty-five vines of each approved Burgundy clone at five dollars each. Most growers took twenty-five vines of whatever they were most interested in and planted them for evaluation, which might take years. Cal Knudsen was more interested in Chardonnay, and Robert Drouhin was more interested in Pinot Noir. What nobody knew at the time was that we were blowing up (rapidly increasing quantity of plant material) different Pinot Noir clones than what had been approved. It was indeed the Wild, Wild West.

The other major concern in transporting grape species from one geography to another is phylloxera. This is a dreaded disease caused by tiny insects called louses that burrow into the roots of the grape plants underground, sapping their vitality. From the soil's surface, they are invisible. All the damage is inflicted underground, which ultimately harms a vine's output and turns some leaves brown. The only way to address the problem is to rip out the vines and replace them.

Phylloxera originated in the United States. Accounts differ on how it spread to France in the 1850s. I believe it came from the Eastern United States and not the Californian frontier as some others argue. But wherever it originated, it was deadly. By 1900, two-thirds of all Vitis vinifera vineyards in Europe were destroyed. In France, the disease spread from the Mediterranean to the Champagne region, which today is an eight-hour drive—an enormous area. It was a major trauma for French vintners, and the memory of what happened is etched onto the minds of everyone who works with grapes around the world. The French had to replant their vineyards with American rootstocks, which were tolerant to phylloxera because they had co-evolved with it.

Here's where the plot thickened. There was a section of Pinot Noir grapes at the Fuqua Vineyard (near my own vineyard) that Argyle was leasing. The section was weak, meaning the plants were not producing enough grapes. Brian Croser had noticed it in 1989 and mentioned it to me. As the person with responsibility for growing the grapes, I decided to fertilize them in 1990. But they proved to be weak again that growing season, and Croser suggested I get some academic experts to look at the block.

I called Ken Brown, the Oregon State University's agricultural extension agent. There were not that many vineyards back then, so the extension guys knew where to go. I didn't bother to show up when he came to the vineyard. I didn't think it was important. He took samples and took them back to the university.

The next day, Robert Drouhin and I were on Worden Hill Road going to look at a section of Knudsen Vineyards that he was buying grapes from. I got a call from Ken Brown.

"It has phylloxera, Allen," he said.

I just about drove off the road. It was shocking. I had the only phylloxera within six hundred miles because I had responsibility for the vineyard. Ken Brown had a responsibility to report it to the Oregon State Department of Agriculture.

The state agriculture guys walked into my office the next morning. They were constructive, but they wanted to talk *now*. They wanted to identify the problem. The burning question was, is there more? They had a relationship with the state Department of Forestry guys, who had a plane. They proposed to fly over the vineyards and take pictures. Then they would show me the pictures and see if they and I together could identify more phylloxera. They would start with the vineyards I managed, then spread out. It made sense. The thinking was that it could be spread on muddy boots or dirt carried by a tractor traveling from one vineyard to another.

By now, suspicions about extra-legal clones were spreading. There were whispers. "What's Holstein up to?" I obviously was in a very exposed position. If it turned out that I had blown up illegal Pinot Noir clones, and the result had been the introduction of phylloxera into Oregon, my career would have been over, not to mention the broader consequences.

The next day after the state agriculture people visited me, I was summoned into the director's office in Salem. This was the director of the whole department. That meant he was the top agricultural official in the state.

He sat me down and said, "Mr. Holstein, we want to write a quarantine order on every vineyard you manage."

That obviously would have been disastrous.

"Wait a minute," I said. "You can write any order you want on the site where we found it, but the other sites are off limits." I had anxieties about lawsuits, property values, and other unpleasant possibilities.

"Mr. Holstein," he warned, "we have the authority to remove the vineyards if we want."

My head still reeling, I got back to the office, and a TV crew was waiting for me. I ducked and weaved, trying to avoid them. But the producer said, "Wait a minute—we have a right to do a story on this."

"Fine," I replied. "And I have a right not to be in it!" But the news got out.

When I told Cal Knudsen about the phylloxera, he tried to reassure me. "Allen, nothing is ever as bad as it first seems." It didn't work.

This was August 1990. I had maps showing the spread of phylloxera in France in the 1800s before the age of travel. If the spread in Oregon mirrored the spread in France, we were done. Brian Croser explained that there was a big difference if this is a five-year problem versus a twenty-five-year problem because of the expense of ripping out vines and planting new ones. Nobody knew at the time. Nobody knew where, if anywhere, it also occurred. Nobody knew how it got there. I was right smack in the middle of it. Fortunately, the state agriculture guys started flying, and we agreed there was no more phylloxera in the other vineyards I managed. Whew!

A couple of weeks later, they showed me an aerial picture of a Dundee Hills vineyard that had a hole (or dead spot) in it. It was a positive identification, a location that I had never stepped foot in. Later that summer, they found a third location much further south, near Eugene. There was no connection between the sites. I was completely in the clear. I came to believe that phylloxera had been hiding in Oregon all along. Maybe it had come in on mail order vines to backyards.

But things didn't turn out that well for the Californians. By 1993, it was clear they had a real problem. Rollin Soles, the winemaker at Argyle, and I had an association with the vineyard guys at Mondavi in Napa. They were a much larger company but had a group of smart vineyard people from different areas who were always trying to raise the quality game. Even though our environments were different, we would visit each other to see what was going on. We used to bitch that they would see what we were doing in Oregon and then take it back to California, bragging that they had developed the idea.

Napa was having a crisis with bad choices made decades before. At some point, it was felt that AXR rootstock was a good choice for Napa. The problem was that AXR had vinifera in its parentage, leaving it susceptible to phylloxera. It took decades, but breakthrough infestations began to appear. AXR began to fail, and a massive replanting program would be required.

We saw it firsthand. Rollin Soles and I had an appointment with Tim Mondavi, the son of Robert, who was the family member in charge of the Mondavi vineyards. We went out with him to meet his vineyard guys and get updated on current events. It felt like this was the only opportunity the real vineyard guys had to communicate bad news. They had an audience. One of the Mondavi gods was descending to be with them.

The top vineyard manager would mention a block, or section, of grapes and ask for updates.

"Gone," came the reply. The vines would have to be replaced. The phylloxera was killing them.

The top guy mentioned another block at another vineyard.

"Gone," came the reply.

It went on like this for ten minutes with managers with responsibility for different blocks and different vineyards offering up devastating report after devastating report.

I couldn't understand the scale of damage because I didn't know how large each block of grapes was. But Tim turned white. That told me everything I needed to know. Outsiders like Soles and I should not have been present to learn about such a crisis, but we were. It might have been a defining moment because Mondavi had to go public to raise money for replanting. (Ultimately, the family-owned company had to sell itself.)

Overall, Napa was going to have to embark on a massive $500 million program to rip out and replace vines. The Napa guys got wind that we had a collection of the French clones, and a giant sucking sound appeared. They wanted a source of new clones to replant their vineyards. They never asked about a paper trail as to its authenticity. Just pack it and send it. They had no other source for the Dijon clones.

We flooded them with vines. We packed entire semi-tractor trailers from top to bottom and front to back with vines and sent them south on I-5. It was a boon for Drouhin and Knudsen. Drouhin had us plant the first rootstock blocks in Oregon, and we sent those also to big commercial nurseries that did the grafting.

Much later, I got called into the Oregon Department of Agriculture with some colleagues after they figured out we were trading in contraband. They threatened to turn us over to federal authorities. But the horse was out of the barn by then. Oregon and France had teamed up to contribute to rebuilding California's North Coast, including the Sonoma and Mendocino regions, over a period of ten years. To this day, almost no one understands the real story. Eventually authentic versions of the clones worked their way in the supply.

This pictures shows the extent of Willamette mite damage in 2002 at DDO and Archery Summit. It seemed like they came out of nowhere.

NEW ARRIVALS IN THE 1990S

- Archery Summit was managed by Gary Andrus, who also owned Pine Ridge in Napa. He was a flamboyant guy who talked big and was the first winemaker to sell a bottle of Oregon Pinot Noir for a hundred dollars. Andrus also bought the Fuqua Vineyard adjacent to Holstein Vineyards as well as an undeveloped plot right next door to Drouhin. So they we were my neighbors on two fronts.

- Ken and Grace Evenstad bought a hilltop across from Drouhin in the late 1980s and began building a mansion on it and planting a vineyard. They had no previous experience in wine production, but they certainly had the finances to make the right investments in land and facilities. They got bit by the bug after having sampled one of Ken Wright's wines in a restaurant, and they contacted him. He made their wine at his winery in McMinnville and later in a building in Carlton where Ken Wright Cellars and Domaine Serene co-existed during the early '90s.

- Willamette Valley Vineyards may have started in the 1980s, but in the 1990s they perfected a system of attracting small investors to their publicly traded company. This approach allowed the investors a certain amount of vanity because they could claim they were part owners of a winery! It also put countless ambassadors on the street talking to their favorite restaurants about "their wine." Willamette would grow to be one of the largest in the state.

- King Estate began in the early 1990s. It was and is owned by an Oregon family with no prior experience in the business but a passion for it none-theless. They were previously invested in some land outside of Eugene and at an elevation that I wondered about at the time. It was off the beaten track, but they built a showcase winery and vineyard and developed that end of the valley.

- Lemelson Winery began in the mid-1990s under the leadership of Eric Lemelson. He was an environmental activist and son of a patent lawyer. He was successful in acquiring quite a number of ideal vineyard properties.

- Bernie and Ronnie LaCroute began calling me in the 1980s. Ronnie explained over the phone that Bernie, who was French, had worked for Sun Microsystems in Silicon Valley types and now wanted to start an operation in Oregon. They called on me at Knudsen Vineyard to investigate the new spacing and irrigating of vines. By then, it was clear that certain aspects of it were successful. They ended up adopting it when they began Willakenzie Estate near Yamhill. They later bought a property immediately to my east in 1999 and named it Jory Estate.

- I met Bill Stoller in 1992. He had grown up in the Dayton area, and his uncle still owned a turkey farm in the Dundee Hills that had fallen on hard times. In 1993, it went into bankruptcy and Bill bought the two hundred-acre place in that process. We started talking in 1994 about my developing it under the Argyle vineyard management umbrella with Argyle taking most of the fruit eventually. Stoller was part owner of Chehalem Winery, and he wanted a majority of grapes to go there. But we began planting for him in 1995 using clones and techniques from previous Dundee projects and would develop the property into Dundee's largest vineyard when we left in 2011.

- Mike Etzel started Beaux Frere. He was a hard worker and initially had to raise pigs to subsidize the costs of developing the vineyard. He went on to achieve some of Oregon's highest scores.

- Doug Tunnell was a CBS foreign correspondent. I remember watching him broadcast from Beirut, Lebanon, reporting to Dan Rather. He was born and raised in Oregon. As his journalism career matured, he was drawn back to Oregon to start what would become the very successful Brickhouse Winery.

- Paul Gerry showed up at Domaine Drouhin one day with his wife, Elaine, wanting a tour and tasting, which was not in my wheelhouse. Steve Dorner was with them. Cristom Winery was in the process of birthing.

- Ken Wright had difficulties with his partners in Panther Creek beginning in 1992. He and his partners decided to sell it and go their separate ways. Ken called me one day and said he'd found a buyer for Panther Creek, but one condition was a source of grapes. Would Knudsen Vineyards sell

Panther Creek some grapes for a period so the new owner could get established? We did it. That allowed Ken to start Ken Wright Cellars in Carlton.

- Although Bergstom Winery did not get rolling until the 2000s, Josh returned from studies in France with his Burgundian wife Caroline and began making wine in 1999. Bergstrom consistently scored high and contributed to Oregon's increase in stature.

- Cameron actually started in the 80s but did not get traction until the early 90s. Like Ken Wright, owner John Paul worked the Portland market himself and developed a following.

Notes: With the exception of Ken Wright and Gary Andrus, no one of the above had any commercial experience with grapes or wine as did Drouhin and the Argyle group. They mostly had other businesses and so were not as dependent on success as the pioneers. But some of them were billionaires and brought considerable investment and business experience. They knew from previous experience to hire good people. Others brought their own brand of passion for wine and helped lift the game for Brand Oregon.

Both art and science are involved in managing grapes and deciding when to prune grape vines and when to harvest them. I was more scientifically and mathematically oriented.

Part of my job was crop estimation, as previously explained. From 1991 through 1994, we all got better at it because of a technique developed by Steve Price, a professor at OSU. He determined that if we could establish the average weight for a cluster of grapes fifty days after bloom, we could predict what tonnage we would have at harvest, hundred days after bloom. It was a real breakthrough.

I adopted his technique and commercialized it. If sugar levels were too low, that might mean we had too many grapes on the vines for reaching optimal ripeness. Or if after testing cluster weights and discovering we were going to have too many grapes for the amount of harvesting and processing equipment that was available to us, I would send crews through a vineyard to cut off a certain percentage of clusters, dropping the grapes to the ground.

One of the biggest decisions in making wine is, when do you harvest the grapes? In the early days, there was an emphasis on the sugar percentage or Brix. So there was an annual exercise to sample the grapes and then test them. If they were over a certain threshold, the grapes were harvested.

Both Croser and Drouhin brought the benefit of their experiences to Oregon, and that improved the collective practice. One kind of mistake was to conclude that the sample that you took was representative of the field as a whole. In this case, the grapes are harvested, and when the wine is in the tank, the winemaker finds out that his sample was too optimistic. This can happen when the sample is too small. If the sugar is too low, then the quality is lower than expected.

Croser had developed a technique that we showed people we bought grapes from. It included a protocol that required taking samples every ten rows, not just ducking into one corner of the field and drawing conclusions. These larger samples were processed in a device that treated the grapes much like a press would. We got much better samples this way and made better decisions.

Robert Drouhin, being French, challenged the emphasis on sugar levels and lab analysis. He wondered, what did the grapes taste like? He would look at the lab analysis, but he also wanted to see the seeds inside the grapes. He would eat a grape and spit the seeds out into his hand. He knew there was a correlation between ripeness and seed color. The browner the seed was, the riper it was. He told me that he liked to see the vine ripen at the same time as the fruit. That meant he wanted to see some sense that the vine knew fall was coming; maybe the leaves had a little yellow in them. It seemed he was making choices that were almost instinctual.

In July or August of any given year, the Drouhin camp would often call and ask me when harvest would be because they wanted to buy their plane tickets from France. Harvest is generally 100 days from the time of bloom. I knew when bloom was, so I could give them a general date, but weather could significantly influence the actual harvest. All it would take would be two or three days of ninety-degree weather, and we would have to accelerate the harvest. So one year as the time drew near, I sent sample results to France ten days before the Drouhin family's scheduled arrival and told them the forecast and said I thought the grapes would be ready. They wanted to get off the plane and get up the next morning to start harvesting.

One thing I learned about myself and about vineyard guys in general is that we tend to be impatient and unqualified to make harvest decisions. It is a little like having a baby each year and giving it up for adoption. The winemaker is the adoptive parent and needs to make the decisions about bringing the baby into the world. The biological parent just wants to get it over with.

That year, I told Robert Drouhin that I would have a crew and start the harvest the morning after his arrival. The numbers all said it was time, but he could call it off if he did not like the maturity. Still jet-lagging, he strolled out to the vineyard around 8:00 a.m. He tasted the grapes off of the vine.

"But All-aan, they are not ripe!" he said.

He was relying on years of experience and a kind of artistic sensibility. I called the harvest off until he was satisfied the grapes were ready.

Around 1998, Rollin Soles surprised me with a surprise bottling from my vineyard. He labeled it "Cowhouse," an obvious play on the name Holstein. It was well-received although in those days my vineyard was at the upper limit in terms of elevation, and some years it did not get fully ripe. In those years, the grapes went into sparkling wine. My vineyard continued to be leased by Argyle during the 1990s, and it was a good deal. I was paid a salary to take care of my own vineyard using their equipment! And I loved having Cowhouse on the market.

THE BIRDS ARE COMING!

BIRD CONTROL GRADUALLY BECAME MORE of an issue in the 1980s and 1990s as more vineyards were planted. If the migratory robins and starlings started arriving before we started to harvest, we had real troubles. The birds would swoop in and eat the grapes right off of the vines.

At first, we had propane cannons, which were sound-making machines intended to scare the birds. But they sat on the ground and had limited range. And if we forgot to turn them off at night, the neighbors got upset.

After a while, I figured out that I could apply for an agricultural fireworks permit with the state fire marshal's office. That allowed me to go to wholesale fireworks supply companies and buy whatever I wanted. I figured out that we needed detonations up higher in the sky. I could buy M-80s, each one of which was equivalent to perhaps twenty firecrackers. I would stuff them into rockets and put the rocket cap back on. Then I built rocket launchers by nailing irrigation piping to a pallet. Then I laced the rockets into the pipes and attached a slow rope fuse from one rocket to another so I would not have to be too close when all hell broke loose.

I would do that at intervals around the vineyards. It was quite a show.

We also had pistol-launched cartridges made in Germany for the same purpose. They had bangers and screamers. There were years where all I did for ten days before harvest was to light up the sky with this gear. That all changed after the 9/11 terrorist attacks. Federal authorities took over the regulation of fireworks and required everyone to get the same permit one needs to buy dynamite, which is about twenty times more powerful than M-80s. Once again, I had feds with badges in my office, displaying little or no sense of humor.

CHAPTER FIVE:

Oregon's Wine Capital Takes Shape

THE WILLAMETTE VALLEY STARTED ATTRACTING large corporate money as well as major financial institutions in the 2000s. This new wave of capital, which was different from previous investments made by wine families and wine-making companies, created new opportunities but also new challenges. One was that big companies do not display much sentimentality when it comes to dealing with people. Decision-making becomes more remote. Another was that as more and more vineyards were planted, we came closer to having a "monoculture," much like Florida orange growers have because their orchards are so close to each other. That meant disease control, and the use of chemicals, became hotter issues. And the town of Dundee had to decide whether to embrace its role as the capital of what was becoming a world class wine region. I'll talk about these three themes—corporate money, the environmental debate and Dundee's emergence—in separate sections to make them more understandable, breaking from strict chronological order.

1. THE CORPORATE GAME

Brian Croser had spent ten years with the Argyle and Petaluma brands. Drouhin's import company, Dreyfus Ashby, was selling Petaluma and Argyle in the US, but sales were not overwhelming. And Croser was ambitious and wanted more. He never told me this, but this is my interpretation: he decided he needed more Australian brands to attract more attention and commitment from distributors. Some of these brands would be at different, cheaper price points than Argyle and Petaluma. To finance the strategy, he would take Petaluma public in a way that left him with a large degree of control. The majority of shares would be held by a tight group of directors whom Croser trusted. He started showing up in Dundee with new people talking about "share price." I had no idea what that meant. But it was all about money.

At the same time, Croser was having challenges with his partner in Argyle. The partner was in it for money and wanted to know, when was he going to get it? Going public was Croser's way to buy him out and get rid of him. In the process, Croser and Knudsen began talking about whether Knudsen should sell his stake in Argyle. If Petaluma was now going to become a public company owning Argyle, why would they want to be partners with some old guys from Seattle? I did not want Knudsen to sell because he was part of my security blanket. Things were changing but in ways I had no idea about, much less control. Knudsen would remain a supplier, and for a while he remained as chairman of the board. The deal went through.

Concurrent with all of this, Croser developed a relationship with Chateau Ste. Michelle, a powerhouse winery based in Washington state. They had distribution horsepower that Croser recognized. They came up with an agreement to sell one of Croser's newly acquired Australian brands if Croser would ramp up production. He did. At Argyle, we hoped we would be swept up by a bigger company that would sell more wine. We had Lone Star vineyard coming into production in the early 2000s.

Things did not exactly work out as Croser had hoped. I was in Hawaii for Christmas of 2001 and got a call while getting out of the water from Croser. I never got calls from him at Christmas. He sounded emotional and basically said he had lost control of Petaluma and by extension, Argyle. I went back in the water.

It turned out that an Australian beer company called Lion Nathan had swept in and seized control. The company had been in the news during the 90s. They had been in court in New Zealand, trying to force a winery to sell itself to them. They had been very aggressive. Lion Nathan was owned by Kirin Beer of Japan and even mighty Mitsubishi, one of Japan's largest industrial groups, had shares in Lion Nathan. The Lion Nathan board thought they were losing market share to wine. So they decided to get into the wine business. One of their merger and acquisition experts walked into a wine shop in Sydney and identified which wine was selling at the highest price and, at the same time, was made by a publicly traded company. It was Petaluma. Lion Nathan offered something like a 40 percent premium over what Petaluma shares were valued at and Brian's close associates on the Petaluma board agreed to it. Done, just like that. Now Argyle, along with Petaluma and the other brands Brian had developed, were owned by a giant beer company.

Officials from Lion Nathan began showing up in Dundee. One guy was Peter Cowan. He was the CEO and an energetic guy. I asked him about his experience with wine, and he told me he'd previously sold and distributed Tampax in China. Great, I thought to myself. His mindset was moving boxes from point A to points B, C, D . . . There was no "sense of place" for him. One of his first acts was to axe the distribution deal with Dreyfus (because Dreyfus did not want to handle Australian brands) and sign a distribution agreement with a Chicago company that walked like the mob and quacked like the mob. This distribution deal would be for Argyle and all the Australian wineries. A slow fuse had been lit.

On another front, I had met Bill Hill, a flamboyant Napa winery owner. Somehow, he had convinced the California Public Employees' Retirement System (CalPERS) that buying vineyard land up and down the West Coast was a money maker. CalPERS was, and remains, one of the largest pools of capital in the world because it is a pension fund for California state employees.

They started in California but also invested heavily in Oregon in the early 2000s. At first, I think the strategy was to buy large parcels, plant them, and then split them up into smaller "lifestyle properties with vineyards" that would attract wealthy retired doctors and dentists or else Silicon Valley types who had hit it big. That way, they could capture the value for multiple homesites on pieces of land where there previously had been none or only one homesite. But that strategy did not work out well because of the recession that followed the Global Financial Crisis, 2008–2010.

So they decided to cultivate the grapes and sell them to Oregon wineries. They had several hundred acres and became a substantial supplier to up-and-coming producers that otherwise could not have afforded a vineyard. That allowed those producers to make and promote their wine and by extension Brand Oregon.

They hired Bob Bailey of NW Vineyard Management do the planting in Oregon. He did a good job, using spacing and clones modeled after the French Opus One experience. The management decided to give a tour to local Oregonians, and someone asked Bailey, "Where are you going to sell all these grapes?"

Bailey answered by asking his own question: "Where are *you* going to sell all of *yours*?"

His implication was clear: the new clones and system of spacing of vines first introduced into Napa by the French at Opus One and adapted by Bill Hill would beat out many other growers. CalPERS spent vastly more money on developing vineyards than I would have or for that matter any Oregonian would have. They went overboard, as if doing so was a selling point.

Then, in 2006, Chateau St. Michelle, which was owned by a big publicly traded tobacco company, Altria (the former Philip Morris), bought Erath Winery. Ste. Michelle became the largest wine company in the Pacific Northwest and was big time corporate. (Altria sold Ste. Michelle in 2021 to a private equity group, Sycamore Partners, which was based in New York, further solidifying the trend to distant ownership interested primarily in dollars and cents.)

There may have been more, but these three examples—Lion Nathan, CalPERS, and Ste. Michelle—demonstrated how larger amounts of money were being invested by companies that never would have considered locating in Oregon a mere ten years earlier. Oregon was a relatively small part of their global business models. Everyone who had come to Oregon in previous waves was "all in," meaning totally committed emotionally. But the big guys expected their Oregon operations to conform with their business models and profit expectations. Passion was conspicuously absent.

But a symbiosis began to appear among producers. Even before the sale to Ste. Michelle, Erath Winery had begun to sell on price, meaning offering entry level wines at prices that distributors could digest. There were more like Erath, such as Willamette Valley, Dobbs, and Duck Pond. This dynamic enticed consumers to pick up affordable Oregon wines and then hopefully trade up. At one point, some 60 percent of Oregon wine by volume was sold by six or seven producers. Other producers, and I tended to be more associated with them, were responsible for the lion's share of the awards and flattering press, which made some consumers want to go pick up a more expensive bottle. In my travels, I've concluded that 80 percent of the producers and winemakers in a particular wine region are riding on the coattails of the 20 percent who are truly dedicated to quality. That is true in Oregon as well.

The larger wineries selling at introductory prices naturally wanted to control their costs. Some of them looked to southern Oregon where it was warmer and they could achieve higher yields and lower costs, which they passed onto the bigger wineries. Most often, these wineries would blend it with Willamette Valley grapes. Southern Oregon had different soils. Rarely, if ever, was there a stand-alone southern Oregon winery that got a big score for their Pinot Noir. It was the Willamette Valley that developed the reputation of Oregon. But these big wineries could truthfully label it "Oregon Pinot," and consumers in other parts of the country could not make the distinction between "Oregon" and "Willamette Valley."

The big ruckus occurred when California wineries started to buy southern Oregon grapes. California has different labeling laws. A wine has to be 90 percent Pinot to be labeled as such. So, they could blend it with 10 percent of California grapes from any region they wanted and call it "Oregon Pinot." We could tell large amounts of grapes were going out of Oregon. A survey was done every year, and there was a big gap between what growers said they produced and what wineries said they crushed.

Meanwhile, Willamette Valley wineries continued to triumph in the wine media. In its day, Argyle produced wines that ranked in the Wine Spectator's "Top 100 Wines of the Year," not just once, but three times. The kicker is that they were for three different types of wine—a sparkling (which is like Champagne), a Pinot Noir, and a Chardonnay, all under the watchful eye of Rollin Soles. When I thought about it, I realized that probably no other winery in the world has done that. Napa Valley does Cabernet and Chardonnay. The Burgundy region does Pinot Noir and Chardonnay. Champagne does Champagne. Nobody wins awards or recognition for three distinctly different types of wine.

It took a while for Lion Nathan to realize that they owned us at Argyle, but bit by bit they began to assert themselves. They had an office in Kansas City. The playbook for these corporate types was to consolidate all of the back-office functions of the various companies they had acquired in one location. They also owned a winery in Sonoma named MacRostie as well as a marketing company. I was the trustee for the Argyle 401k program and was the one who started it. But Lion Nathan moved that function. Health insurance was now their domain. I had been a signer on the Argyle checking account, but no more. Our existing office staff either quit or were fired. Lion Nathan trotted in a human resource person. Her last name was Botter. I called her Bother. We got in a fight once when I fired someone. Like I was supposed to get her permission. Rollin Soles and I wondered if we needed to write an employee manual to conform to the policies of our new masters. No, we concluded. If we had one, we would have to follow it!

In the mid-2000s, we could see the end was in sight for finishing the planting at Lone Star Vineyard, which we had started in 1997 in Eola-Amity Hills. Being thirty minutes south of me, it was out of my Dundee footprint, but I had spent ten years planting it. Now it came down from the corporate strategists that we were going to need still more vineyard land. I started sniffing around in 2007. I had my eye on a place in Eola, a couple of miles as the crow flies from Lone Star but higher in elevation and 100 percent Jory soil.

I did a little work and learned that it was owned by Weyerhaeuser. I knew better than to call their telephone number in Washington, explain I was a neighbor in Oregon, and ask, who could I talk to? So I called Cal Knudsen, who used to work there. A few days later, he called and said, "I ran into George Weyerhaeuser on the golf course, and I got a contact for you."

So began the process of purchasing Spirit Hill Vineyard for Argyle. As one would expect, Weyerhaeuser had a real estate division that oversaw twenty-six thousand acres in Oregon and Washington. They were pros.

They asked for the name of an appraiser who specialized in vineyards. We gave it to them. They engaged him. Then, they announced that they were going to auction the piece of land off and gave notice to all the neighbors that they wanted to sell. They were playing us off against the neighbors. In our written bid, Rollin Soles cleverly said he would pay one dollar more per acre than anyone else. They were a little dumbfounded by that. But we got it, a 180-acre piece where we would plant 135 acres. We began planting in 2008 as the financial crisis was unfolding. But money was no problem for Lion Nathan because it had such deep pockets. We finished in 2015, all with high density. It was a capstone achievement.

START-UPS IN THE 2000S.

PEOPLE WHO HAD DEVELOPED CONSIDERABLE expertise and extensive networks of contacts while working for others began thinking about and planning their futures in a world that had gone increasingly corporate.

- Rollin Soles began planning life after Argyle, and we planted a vineyard for him in 2002. He eventually launched the Roco Winery. He sold a majority ownership in it to Santa Margherita USA in early 2022. It was part of a large Italian company.

- Bill Hatcher and his former wife, Deb, partnered with Sam and Cheryl Tannahill to create A to Z Wine Works.

- Laurent Montalieu left Willakenzie Winery in 2003 and began his own business called Northwest Wine Company.

- Lynn Penner Ash left Rex Hill during this era and began Penner Ash Winery.

- Joe Dobbes left Willamette Valley to begin Dobbes Winery.

Bill Hatcher had guided Domaine Drouhin for over a decade, and it was a clear success story. He felt like family. We were all close. In the ongoing tension I had with the Drouhin family about being a contract vineyard manager rather than a Drouhin employee, Hatcher supported me. The arrangement was working.

But one of the challenges we all had with absentee owners was finding the right balance between authority and responsibility. If we were assuming ownership of certain responsibilities, I always felt I should have the authority to look at the facts on the ground and execute a plan. Absentee owners did not always agree. Bill Hatcher had his own challenges in that regard, and he left Drouhin in 2000 after a long and successful tenure.

Soon enough, the Drouhins hired someone who had an interest in wine but was an entertainer on radio. They did not know him. On the very first day, Robert Drouhin sat us down and looked me in the eye and said, "You are not the story and neither are you," turning to the new hire. "Véronique is." She did, in fact, become the public face of the brand, its ambassador. She was the winemaker too. In fact, she became an important ambassador for all of Oregon wine.

Bill Hatcher and his wife, Deb (since amicably divorced), went on to partner with Sam and Cheryl Tannahill in A to Z Wine Works. Sam Tannahill had worked as a winemaker for Archery Summit. Gary Andrus, the principal owner at Archery Summit who came to Oregon with Napa-sized expectations for price and volume, had an oversupply problem because all his vineyards came into production during the post-9/11 recession. The Hatchers and Tannahills bought the overage. It was high-quality. So they got high-quality wine at modest prices, and that bailed out Archery Summit. Then A to Z sold the wine at "democratic" prices and developed a following. It turned into one of the biggest brands in the state.

I had had a good run with the Drouhin family, but we did have some conflict. One was the conversation about farming practices discussed below. Another was that Robert Drouhin was concerned about depending too much on me when I also worked for Lion Nathan. He gave me a lecture after he learned that Lion Nathan had bought Argyle. He knew, better than I, that that type of corporate control would change my little world. If I have one regret in my career, it is that I did not consider going full time with Drouhin. It surprised me in 2007 when Philippe Drouhin told me they would not extend the vineyard management contract with Argyle after the harvest of 2007, meaning they would no longer require my services. That was it. I did not argue. After the harvest of 2007 and after twenty years, we parted company and remained on good terms. It had been a grand adventure.

2. THE ENVIRONMENTAL DEBATE

The late 1990s and early 2000s were a peak time for me to participate in vineyard tours aimed at wine merchants and other sales people coming from out of state. It was also the peak time of excitement about the breakthroughs in viticulture that we had navigated. Quite often, when I demonstrated our latest viticultural innovation, someone from Atlanta or Houston would raise their hand and ask, "But is it organic?" I began receiving more and more inquiries from sales people as well as from winemakers and owners. Philippe Drouhin, son of Robert, and I had epic arguments about it. He had joined a group of Burgundian producers who were sold on the idea that "biodynamic" practices produced better quality grapes and were better for the environment. I regarded his biodynamic practices as a kind of voodoo. They advocated planting vines depending on the phase of the moon, for example. I kid you not. But it turned out Philippe had his name on the sign out front!

My baseline position had long been that the wise use of chemicals made sense in terms of controlling weeds and pests. If we did not use any chemicals, we would have to spend far more money on labor, equipment, and fuel to operate the machinery. Never were the go-organic proponents willing to tease apart the issues and discuss them in terms of worker safety, consumer safety, and environmental safety. And never mind that conversion to completely organic practices involved higher costs—and we were constantly trying to avoid that.

In hindsight, it was the intersection of the politics of selling and science. Chemicals sprayed on grapes are not passed on to consumers because the grapes are crushed and only the juice is used to make wine. Federal authorities have established thresholds below which the wine is considered safe. One would have to drink fifty gallons of wine a day for a year to exceed the threshold.

I tried a two-track approach. I agreed to learn more about organic practices and apply them at Stoller and Drouhin on certain blocks of vines.

Using what would be described as a "sustainable" spraying program, we would run the tractor through a particular block of grapes eight times a year. The organic crowd might choose to run tractors through eighteen times. I argued with them, saying "Ten or twenty years ago, when faced with the question of 'Do I use a chemical or do I use more diesel?', many people would have been happy to use more diesel since that was not considered a chemical. But as time passed, we learned more about the upstream and downstream effects of using more diesel. Diesel had a chemical impact where it was produced and transported from. And then downstream, the use of diesel contributed to climate change.

Experimenting with the organic approach also meant we needed to double the number of tractor drivers. And we still experienced problems with mildew, meaning that our crop level was lower. All these arguments fell on deaf ears. I got the fish eye when raising the above question.

Another part of my hedging strategy was the Low Input Viticulture and Enology (or LIVE) initiative, which began in the 1990s largely as an educational organization assisting new growers. Then it morphed into a broader "sustainability" group. I could see that my continued heel-digging could result in my career being prematurely terminated. So I led Argyle, Knudsen, Drouhin, and Stoller into the group, providing much needed dues and credibility. I got myself on the board as treasurer. This was in 2005. It involved third party inspections of vineyards and wineries and had some level of credibility although critics called it "organic light." LIVE approved a spraying program involving synthetic fungicides, for example, which was controversial among the purists. We hired Chris Serra to be the program director, and he did a good job.

The program was expanded to certify wineries as well as certifying wine made from certified vineyards at certified wineries. We expanded to Southern Oregon as well as to Washington. At one point, certified vineyards made up almost 35 percent of total Oregon acreage. Nobody else had a program like that. Sometimes there were workbooks or guidelines, but nobody had an inspector visiting the vineyards to ensure compliance like LIVE did. Adding to its credibility in my view, LIVE made judgments about "the whole of farm," meaning an entire property. The organic guys allowed inspection of just one production block, but not the rest of the property.

LIVE allowed the use of Round Up, the weed control substance, and that also triggered controversy. The pure "organic" practices used more diesel and arguably caused more damage to the soil. But it did not matter. The messaging jury had decided. (A California court case years later, in which a worker became seriously ill after using Roundup, resulted in the industry halting the use of the chemical.)

Throughout the 1980s and 1990s, we prided ourselves for not having any insect problems and therefore did not use insecticides. From the standpoint of applicator safety, this was a class of chemicals to be careful with. Even so, when I was out in the vineyards with the Drouhins, sometimes they would turn over a leaf to examine it. Then they would say, "You have spiders." They had had experience with spider mites in Burgundy. I thought to myself that so far, the spider mites had not caused economic damage, so why worry about them?

But in the 1980s and 1990s, plantings were scattered, and as such there was no monoculture. But as time passed and more parcels were planted out, the monoculture effect began to appear. At Drouhin, we had become concerned enough about the mites that we had a consultant come every week and do counts in a systematic way and report them to us: so many mites per leaf.

I was stunned in 2002 when the counts went from ten mites per leaf to a hundred per leaf in a week or two. As a result, Drouhin's vineyards began to turn red and yellow in late August at a rate I had never seen before. The concern was that if the leaves were not productive, then the grapes would not get ripe. It was a very warm year, and maybe the carbohydrates from the roots had contributed to a very ripe vintage, but alarm bells were going off about the spider mite problem.

We had the university people come out, and they took samples of the mites and took them back to be identified. The initial answer was that it was a French species and not native to the United States. That would have been seriously bad news because it would have implied sloppy and perhaps illegal practices. But in short order, the university experts corrected the identification—it was the Willamette mite and therefore a native species.

After harvest, I went to confer with the experts at OSU. An expert there asked, "Do you spray sulfur?"

I said we did, to control mildew.

"Well, you are killing the predators," meaning the other insects that ate this particular species of mites. Apparently, there was a whole universe of ecology going on that we had no idea about. There were local predators but something had caused the balance to tip—maybe our sulfur use. We did not want to use insecticides, so how could we tip the balance back in favor of the predators?

The OSU guys knew a blackberry grower whose farm had tons of a local predator known as T. pyri, another genus of mites that preyed on Willamette mites. I drove over to Silverton with a bunch of grocery bags and, with his permission, went through grabbing leaves with the hopes they would contain the predator. It worked. It took time to increase the numbers, but we were able to reestablish the population of our preferred predators.

In 2003, I had a similar shock at Stoller Vineyard. This time it was in the spring, and the emerging shoots began to look like popcorn. The shoots were not expanding normally. We were stumped. We began consulting other parts of the world, and the suggestion came back that it was rust mites. These were much smaller than Willamette mites, and one needed a microscope to see them. What was the solution?

Spray sulfur! At the time, I felt that it did not fit the pattern of a normal insect problem. It was worse on young vines. A young block could be side by side with an older block, and the young block displayed symptoms while the older block had none. Yields were impacted mightily as the mites would chomp on expanding clusters.

At first, it was mostly at Stoller's, but then it began appearing in other vineyards that we managed. It seemed to follow us around. Other vineyards had it but did not recognize it. This was more than twenty-five years into the Oregon experience, and new problems were emerging as more vineyards were planted.

I thought there was something attracting the mites that we should fix. I still do but never could put my finger on what it was. So we eventually moved on to treating the symptom of the problem rather than the cause. There was a window of opportunity in April, before Pyri (the predator) was out, when the rust mite was vulnerable to sulfur. That was when we needed to attack. We had to reduce our pounds per acre per year of sulfur for mildew control and depend more on mixing sulfur with other materials.

Despite the pest and environmental issues and rain that hit the 2005 harvest, every other crop in the 2000s ripened beautifully and contributed to Oregon's rising boat. The movie "Sideways" came out at about this time, teaching people how to pronounce the name Pinot Noir. We were making headway.

3. DUNDEE AS WINE CAPITAL

Dundee was not much of a town when I first arrived. It was located on four-lane 99W, about two miles southwest of the slightly larger town of Newberg. One of the main businesses on 99W in Dundee was a place where hunters could take their game to have the meat processed. There was a shooting range south of town. Some residents worked at a major paper mill on the Willamette River in Newberg. In short, Dundee was a solid blue collar kind of place. Traffic was brutal on weekends because 99W was one of the main routes that Portlanders took to go to the Oregon coast, only about an hour away from Dundee. Crossing 99W on foot could be hazardous.

The only place in Yamhill County that would serve wine with a meal was Nick's Italian Café, which was located on Third Street in McMinnville. The winemakers supported it, all thirteen of the original pioneers. "The heart and soul of the Oregon wine industry was fashioned in the back room of Nick's restaurant by a handful of starry-eyed newcomers," wine critic Paul Greggut would later write.

It took a while for the restaurant scene to develop in Dundee. There was a place in Dundee called Alfie's. It was more for the Lawrence Welk crowd. Dundee also had a breakfast and lunch spot called Scampy's owned by a retired navy guy and his wife. Before long, a hip young couple from Santa Cruz, Alice and Russ Halstead, showed up and bought it. Instead of selling greasy burger gut bombs, she, the leading decision maker, offered a Dundee Dandy hamburger and a Gaston Gold cheeseburger. She and Russ bumped along for a while. People from the wine industry started showing up. She decided to open on weekend nights and serve fine cuisine. This was the 90s. Wine people filled the joint. After a while, she hired some folks to do the cooking, Dave and Tina Bergen, while Alice acted as hostess. Intrigued by what they saw, Dave and Tina moved down the street and started Tina's restaurant.

Alice and Russ later started the much larger Red Hills Provincial Dining, which also offered excellent international quality cuisine.

The big break was when the Ponzi family opened the Dundee Bistro in the late 1990s. It was a hit with the wine community. Others soon started the Red Hills Cafe, which brought some of Portland's strong coffee culture to little Dundee. It took a while, but Dundee was beginning to support the tasting rooms with the lifestyle that international wine lovers enjoyed.

In 2010 or 2011, the Red Hills Market opened, and I thought they were crazy to open in that financial environment. But it served a real niche and packed in the customers, including me.

One time on a Friday afternoon in the '90s, I stopped in the Argyle tasting room to get a bottle of wine for the evening. There was a couple standing there, and I chatted them up. I asked where they were from. They lived in New York, but he had been in Japan doing business and they wanted to reunite, after a long separation, somewhere on the West Coast. They wanted to have some sparkling wine at Argyle and then dinner at Tina's across the street. Wow, I thought, they could have met anywhere on the West Coast. The fact that a couple this worldly and sophisticated had found Dundee reflected the dramatically new face the town and the region was showing to the world. In 2016, Oregon was named "Wine Destination of the Year" by one of the wine magazines, and McMinnville was runner-up in 'best small downtowns' in the country. It was dizzying to the old timers

---

In the early 2000s, I was at the library or somewhere and noticed that the state highway department had posted some photos for a potential highway bypass around Dundee and was seeking public comment. These were actual overhead photos of different locations with the Newberg-Dundee bypass inked in. I noticed that they had one option coming down Red Hill Rd., adjacent to my place. "What? A bypass highway coming down my country road? Not In My Back Yard (NIMBY)." I organized a community meeting to discuss it. Neil Goldschmidt, the former governor, owned a vineyard. He was helpful. He knew the ropes.

It contributed to getting the broader Dundee community to ask itself, "What do we want to be when we grow up?" It was decided that we needed a vision, to both to identify bypass location but also more broadly than that, what was our identity? Most of the vineyards and wineries were outside Dundee's city limits, predominantly in the Dundee Hills west of town, but obviously the town was greatly affected by their presence. The Dundee City government knew some facilitators who would help us achieve a vision statement. They were former legislators. I was appointed alongside a broad variety of locals. I was the only wine guy. Most people in the wine industry did not actually live in Newberg or Dundee.

The visioning was quite a process. The facilitator said upfront, "You'd better decide if you want to be the wine capital of Oregon because if you don't claim it, it will go somewhere else." Of course, we did, I thought. Only some of us did not. Some of us would rather have had a Walmart in Dundee so we would not have to drive ten miles to McMinnville. It took a year or more, but to this day the vision statement hangs from the walls at Dundee City Hall and yes, we are the wine capital with a bypass and a welcoming pedestrian-friendly environment for locals and visitors alike. Other wine towns such as McMinnville, Carlton, and to some extent Newberg also have risen in stature, but not like what happened in the Dundee Hills area.

Another piece of the region's emergence was the creation of American Viticultural Areas, or AVAs, much as the French have their appellations. There had been attempts to create different AVAs before, but consensus on the boundaries could not be reached. In the early 2000s, it bubbled up again when Ken Wright, Alex Blosser, David Adelsheim, and others began to talk about it. We had meetings. Ken Wright talked about what differences there were between the grapes grown in Yamhill and Dundee and Eola. I remembered from my days at Erath Vineyard on Chehalem Mountain that there were differences in soils and geographies and therefore tastes.

The problem was that to be recognized by federal authorities, there had to be a consensus around the AVA boundaries, and they had to be discernible. Boundaries had to be a road or a river or something fixed and recognizable. That meant some producers would be in and others not.

Eventually, Ken Wright and the others worked it out. Once the AVAs were recognized in 2004, I was asked to present wines from the different AVAs to Jancis Robinson, a famed British wine writer, and explain the nuance of each. It was kind of an honor. I was not familiar with the north side of Chehalem Mountain and had to be coached a little bit. But it worked out.

I think the creation of the Yamhill Carlton, Ribbon Ridge, Dundee Hills, Eola Amity, and Chehalem Mountain AVAs was a major marketing development, not just for the wines but also the Dundee region. It suggested we had reached global stature. If you added up the number of cases of wine over the years from Dundee Hills that achieved Wine Spectator scores above ninety, Dundee probably exceeded the other AVAs combined. Whether it was the soil or the extent of major investments, it's not clear. But the Dundee Hills AVA clearly stood above the rest. The highest scores probably came from Ribbon Ridge, although in small amounts.

The seeds of Oregon as a wine destination were being planted during this era. For much of its history, the Willamette Valley was something of a backwater. High-end travelers would pretty much have to stay in Portland because lodging was substandard in Yamhill County, which included Dundee and most of the wine country. So the tourists would have to drive themselves out to the valley, drink wine, and then drive back, risking unfortunate encounters with the gendarmes. Napa was a livelier alternative.

In 2009, the Allison Inn and Spa opened. They had an open house in September, and I could not believe my eyes. It was a world class hotel and spa complex with a highly rated restaurant. It brought attention to the area.

Technology in the form of smart phones and the Internet clearly played a role in the region's boom. Although it took perhaps a decade for the impact to play out, smart phones enabled tour operators to flourish because they could maintain better communication with customers and the tasting rooms and coordinate more visits. Bed-and-Breakfast places popped up. But Airbnb was a game changer. Newberg, Dundee, and McMinnville had an ample inventory of rentals to entice out-of-state travelers. It reached the point that the Dundee City Council debated whether to limit vacation rentals inside Dundee. That never would have happened ten years earlier and was yet another sign that the Dundee area had truly arrived.

Ken Wright did a fabulous job of developing Carlton from a logger's town to a real wine destination.

CHAPTER SIX:

The Critical Role of Immigrant Workers

THE OREGON WINE BOOM WOULD have never happened if it had not been for the mostly Mexican workforce. Much of the West Coast's agricultural sector is heavily dependent on Mexicans and some Central Americans, and we were no different.

When I first started in Oregon in 1980, strawberries were a major industry. They would ripen just after school let out for the summer, which worked for the industry because housewives could take their kids to pick strawberries and make a few extra dollars.

I inherited some of the housewives. They were good workers, but they were not terribly reliable. If a child had to go to the dentist, that was their priority, which was completely understandable. The Latino workers started showing up and began picking the strawberries and working the vineyards. Others came, too. One time, a hippie couple came hitchhiking through and wanted to work in the vineyards. They must have thought it would be a culturally enriching experience. They made it until lunch and quit. The Latino workers were simply better adapted to the work and were more efficient than the locals or passers-by.

Harvests were, of course, the time of year when we needed the most labor. It was back-breaking work. Practices have shifted a little bit over the years, but the basic pattern was that workers had to cut off clusters of grapes, put them in buckets, then put multiple buckets into wooden bins, and ultimately load it all up on small tractors that would then take the grapes to the winery, where the grapes would be crushed. (Forklifts were necessary to lift the wooden bins off the tractor beds.) The boxes were tagged with tickets showing specific rows so we would know where the grapes came from. Cries of "ticketo correcto" (the ticket is correct) would ring out along with other shouts such as, "hay un otro" (I need another bucket).

I tried to keep up with them working the rows. I could never do it, even in my prime. So I adopted a trick where I would work alongside someone in one row, which would put two of us in one row. Then I would switch over to another, then another row. No one could accurately assess how little I was able to harvest compared with the Mexicans.

From a distance, it looked like a homogenous group. It was not. Mexico is a surprisingly diverse place. Some groups hardly spoke Spanish, using instead local Mayan or Indian dialects and seemed to hail from isolated areas. Others came from well-developed agricultural areas and still others from the larger cities.

Few of them were authorized to work in the U.S. We just made up Social Security numbers. One time when I was hiring a man, he gave me a Social Security number with only eight digits. I just added one more. Every now and then, we would get a notice from the Social Security Administration stating that a given name and number I had submitted did not exist. To confuse them, I would return the form with the same name and number, only I would add a middle name initial—as if that would clarify it.

It worked for a long time until the mid-1980s when it started getting harder to find workers. Reagan's asylum trick fixed that for a while. If I wrote letters claiming that an individual had been in the United States for a certain number of years, that was enough to make them legal.

Over the decades, I directly hired hundreds and worked alongside them. I got to know their families and their problems. I felt for them. About half lived in Oregon full time, but the other half took enormous risks crossing the US-Mexico border. They would cross from Mexico to work the agricultural seasons in the United States and then go back across the border in the off-season. I used to hear them tell stories while they worked in the fields about the coyotes, the shady, often criminal middlemen who brought them back and forth over the border. An atmosphere of lawlessness permeated the border region.

One time during harvest I found a couple working in one row, and nobody was working the next row. I told the woman to skip over to the unoccupied row so we would not leave it behind. She looked at me and complied. Later I went back and found her working in the same row as her husband. After a few rounds of this I figured out she was scared to be working alone. It made sense after I visualized what she had been through to get here. I found someone else to harvest that row.

I helped one worker named Mardonio to become legal during the Reagan years. He was a crackerjack viticulturist and spoke perfect English and was well-read, a real philosopher. He got into a little trouble with the local police, and he decided to take a trip to Mexico in his pickup. They found him dead sitting in his pickup on the US side of his border— expertly stitched up after his organs had been harvested.

Gabriel was one of my first tractor drivers. His family worked in the vineyard while he drove the tractor. One year, he wanted to take his family home over the year-end holidays. I made a deal with him: if he went and came back, I would share some of my bonus with him. I never knew the circumstances, but he was murdered while in Mexico.

My interactions were not always so tragic, but they were often deeply personal. I did not have a human resources person to hide behind. I was the HR person. For decades, I did not speak English during the course of the day until I went home in the evening. Such was the intensity of my personal interactions with these workers. When I traveled in other Spanish-speaking areas of the world, people would tell me, "You speak Spanish like a Mexican." Or else they would say, "Where did you learn Spanish? We have not used that word for five hundred years." Spanish is just like English in that regard. People in different parts of the world speak it differently.

In the summer of 1998, I got a letter from Immigration Services that they wanted to audit my records about the work authorizations of my employees. Adelsheim got notified, too. It was like Immigration had a list of wineries in Oregon and picked the first two in the alphabetical order. For every worker, there was an I-9 form that had to be filled out specifying what documents in the worker's possession authorized them to work in the US. They wanted one folder of these forms for current employees and a different one for former employees. I guess they wanted to see if I had fired some people between the time of notification and the time of the audit.

I got a lawyer and bought time, but the advice was "forget it, these guys are the feds." They took my files back to their office and made me squirm. About a month before harvest they notified me that twenty-five out of my thirty-five people were ineligible. I had to fire them.

Some of these people had been with me for years. It was heartbreaking to me. Immigration did not go after them. They just went down the road and got another job. In fact, one of my assistants took the opportunity to leave and start his own company and hired some of the people he had previously supervised under me. I had to improvise to get through harvest and then start to rebuild. I had to be careful, not knowing how closely I was being watched.

As bigger and more diversified companies got involved in Oregon wine, many had the policy of zero tolerance for the hiring of undocumented workers. That spawned a whole cottage industry of "labor contractors," who assumed the liability risk of immigration status, which largely protected the client winery. It is one reason for the growth of labor management companies.

Another problem was health insurance. I lobbied to get major medical insurance for the workers when it was available in the 1990s. But it did little good for them to be able to see a doctor if they could not afford to get the medicines that were prescribed. So eventually I lobbied to get them regular health insurance. Only about 20 percent of the wineries provided it. As middle management jobs began appearing and were occupied by mostly Americans, they asked for top tier insurance plans that employers had to provide to all employees without any hint of discrimination. But it was a trade-off for vineyard workers: do you want a raise or health insurance? They wanted the raise, not the insurance. So, for that reason, it was much easier to use labor contractors that did not provide any insurance.

The stories were rich, and there is another I would like to share. Jaime Cantu came to the US as an unaccompanied minor at the age of fifteen in the 1980s. He told me he had to lie about his age to get his first job in the Dundee area. I hired him at minimum wage in the early 1990s. He proceeded to kick down every door that life put in front of him. First, he became expert at driving the French tractor at Drouhin's. Later, I put him in charge of planting the first vines at Stoller, not knowing how big of a job it would turn out to be. In the process, he turned out to be one of the best viticulturalists in the area. We had a good partnership. I used to say, "His job is to keep the wheels on the truck, and my job is to keep the truck on the road." That is what it felt like.

As my career matured and people began looking around at who would come next after me, around 2015, I thought Cantu was a natural. But the leadership of the big corporate company thought otherwise. The job went to a college-educated white man who had never stepped foot in our vineyards or had relationships with the workers.

In recent years, I watch as the vans full of pickers show up at my vineyard to harvest grapes. As they emerge from the vans, I notice that many are my age (in their sixties) and still doing this. One of the tragedies is that they worked under phony Social Security numbers, which means they contributed payroll taxes to the system. But they would never get anything back from the government.

There are forty thousand acres of grapes in Oregon. More or less, it takes two hundred man hours per acre per year. That translates to four thousand people working in Oregon vineyards that are 95 percent from Mexico and Central America. We owe them an enormous debt of gratitude.

But nothing stays the same. What we pay our workers has had to increase, reflecting the basic laws of supply and demand, and that has major implications. We are mechanizing some aspects of grape growing, particularly harvest. But there is only so much of that we can do. For sure, other wine regions have prospered without the benefit of migrant labor. In the Burgundian model, the individual acreages are smaller, and the owner-operator does some of the work himself and sells a higher percentage of his wine directly to consumers.

The business model of growing grapes and selling them was always challenging in a cool climate where quantities of grapes per acre is low. But with labor prices up, that puts additional pressure on them. Indeed, my contacts in Burgundy tell me that few serious people only grow grapes. They need to have the value added of making the wine, and that means having a winery. That means a bigger investment. Stand-alone vineyard operations (which I once operated) will be challenged. That's bad news for the stand-alone vineyards still in operation. I'll talk about the problems facing some of the smaller operations in the next chapter.

CHAPTER SEVEN:

_____

Into The Future

As the old cliché goes, the only thing that is constant in this world is change. As we entered 2022, it was clear that trends both inside the Oregon wine industry and outside of it were going to force it to continue adapting. Wine critic Paul Greggut in January explained that one piece of the challenge was "the difficulties of transitioning small businesses across generations."

But as he wrote in his blog, there were other factors, too: "It is partly the result of the piling on of obstacles to smaller wineries in particular—vastly increased competition, limited distribution, Covid shutdowns, wildfires, disappearing seasonal labor, and on and on." I've already talked about the labor issues, and I will talk about some of the other issues in this chapter—but save the climate discussion for the final chapter, because it is arguably the biggest unknown.

The transitions at Argyle started in 2013 when Rollin Soles said he would step back from his winemaker's role there to concentrate on his own brand, Roco. He had been quietly building the Roco brand in addition to custom crushing the overage of grapes from Argyle that they could not squeeze through their original facility. His departure from Argyle was the end of an era. To quote another cliché, all good things must come to an end. We had had a great run of bringing fun and science to the wine business. We had worked together for twenty-five years, like a marriage. But the corporate culture under Lion Nathan was becoming, well, too corporate.

That same year, I wrote a "white paper" for the Argyle leadership assessing grape supply. My conclusion was that many vineyards had been removed during the 2007–08 financial crisis, and not enough had been replanted to take care of the demand, which was beginning to increase once again. The Spirit Hill vineyard was nearly all planted, and the bosses were looking for the next source. So Argyle tasked me with the hunt for more land because the assumption was that the supply of grapes for purchase was insufficient.

I knew that the best land near the center of our current operations in the Dundee Hills was going for a premium and that to get into Argyle's lower price range, we would have to go further south. The problem was I had no idea how I would manage another location that far from our existing radius.

To put it into perspective, I once did a calculation and determined that if I put all of our rows at all of our current locations end to end, the resulting row would run from Portland to San Francisco. It makes the vineyard manager a little crazy. In assessing the day-to-day interactions of vineyards with the weather, he cannot be everywhere all of the time. So, he stops in Sacramento, for example; makes an assessment; and then wonders if that assessment applies to Eugene in Oregon. Not sure, he drives to Eugene to see, recognizing that one snapshot of the conditions in Eugene may or may not apply to the other side of Eugene. Assuming it does, he then wonders about the rest of Willamette Valley. By this time, the weather changes in San Francisco. It makes you crazy. I was spending more time with my "nose to the windshield" than I was actually growing grapes. I did not really want any more rows.

CalPERS must have come to the same conclusion I did in my white paper because they decided to gradually sell off the vineyards they had planted more than a decade ago. They sensed opportunity. They wanted to get out of the grape-growing business, and the world had recovered from the global financial crisis. Their decision to start selling was a major development.

This attracted well-heeled companies that needed scale to justify adding Oregon to their existing portfolios. These were all hundred-acre-plus vineyards in full production. Jackson Family Wines was one of the first to purchase one of the CalPERS properties. I remembered meeting founder Jess Jackson in the 1990s and asking him if he was considering an Oregon project. We were always trying to entice big name folks to invest in Oregon. But he said no, the risks were too high and Oregon's yields too low.

Now, however, the future had arrived. The Jacksons went on to purchase existing brands in the name of Willakenzie and Penner Ash. The Jackson family was a juggernaut in the marketing world and would bring more attention to Oregon.

Ste. Michelle bought one of the CalPERS vineyards named Willakia and used it to bolster Erath's production so they would not depend on Erath's grapes and the prices he wanted.

I found out that one buyer of a CalPERS vineyard was Atlas Vineyard Management. I knew one of the principals, Mike Cybulski, from the days of visiting Mondavi's vineyards. Atlas was a big management company that primarily operated in California's North Coast. I knew that Cybulski was too smart to expect to make a big profit from selling grapes. So I made the introduction to the Argyle leadership with an idea to partner with them. It was a breakthrough—Argyle found an additional source of grapes, and Atlas would manage it, not me. The partnership eventually came to be known as the Giving Tree Vineyard.

As mentioned previously, Lion Nathan had purchased a Sonoma winery called MacCrostie. They depended on management companies, but I was still charged with going down to look over their vineyards once or twice a year. I helped them with due diligence on a purchase outside Healdsburg where they later built a tasting room. Specifically, I was asked to help with crop estimation. This was a crack up. They were not accustomed to the exercise either at MacRostie or the vineyards from where they purchased grapes. I was kind of imposed on them, but they were gracious. Once I was in a vineyard from which they purchased grapes. I was out in the vineyard in August selectively choosing which grapes to sample when someone associated with the vineyard wandered by. They asked what I was doing. I told them. They looked at me like I was crazy.

There I was, buying land, cruising Sonoma, and keeping up with five hundred acres of grapes up and down the Willamette Valley. I did not know it at the time, but I was hitting the sunset years. I wasn't as spry as I once had been.

Jackson had too much weight on back of tractor, so I served as ballast

THE NEW WAVE OF WELL-KNOWN WINE COMPANIES SETTING UP SHOP IN OREGON

- Louis Jadot, one of Drouhin's biggest competitors in selling Burgundies worldwide, bought a piece of land right next to Drouhin in the Dundee Hills. It was obvious that Jadot had plans for Oregon, but it was said they would be methodical about it—and they were. They waited years before buying the small Resonance brand from Kevin Chambers.

- The Henriot Group bought Beaux Freres. Henriot already owned Bouchard Pierre and Fils, making it the third-largest Burgundy negotiant to invest in Oregon. They were another company with a worldwide distribution network selling the most respected Burgundies right next to their Oregon wines.

- Bollinger Champagne bought Ponzi winery. Bollinger had been a partner in Petaluma Winery back before it was purchased by Lion Nathan.

- From Napa, the Francis Coppola group bought a second tier winery adjacent to Domaine Serene. And Silver Oak of Napa bought Dick Erath's original vineyard off Worden Hill Road that I had planted in the 80s. It had an expensive home on it. They tore it down to put up a tasting room.

- In 2017, Domaine Serene built a hospitality center that could have been a chateau in Bordeaux, it was so lavish.

The net effect of all the major wine companies placing their reputations on the line by investing in Oregon was profound. Smaller business owners, particularly those with no children, were able to make their exits from the business with their skins intact. The ownership structure of Oregon's wine business was once again shifting. The traditional problem of distribution that Oregon winemakers faced began to ease as big Californian and French wine companies moved in.

One deal in particular caught everyone's attention. Early in 2022, Rollin Soles sold majority control of Roco Winery to the Santa Margherita wine group of Italy. It was their first investment outside of Italy and suggested that other non-French foreign firms might beat a path to our area. Our land prices, although significantly higher than when the wine boom started, were still low compared with other wine regions of the world. "It's a super validating time for us in the Willamette Valley as we've seen recent Euro investments by Bollinger, Henriot and Jadot," Soles told Paul Greggut's wine blog. "Smart high-quality Euro winemakers recognize the potential, the beauty and the authenticity of Willamette Valley wines." He said getting increased sales representation and distribution through Santa Margherita would be a big win for Roco.

It was also a win for the entire Willamette Valley because of the logic behind Santa Margherita's purchase. As Chief Executive Officer Beniamino Garofalo explained to "The Drinks Business" website, the company was selling wine in the United States, mostly Pinot Grigio, through a wholly owned distribution outfit. But because it lacked a winery, under US law it could not have direct contact with American consumers. That's what Santa Margherita really wanted—direct e-commerce, wine tastings, and wine clubs, the very formula that Oregon had pioneered. Our build-out of a tourism industry was another piece of the company's thinking. Oregon is one of the fastest-growing wine regions in the United States, Garofalo said.

As for me, I retired from Argyle in 2017. Lion Nathan tore down the winery operations in Dundee and moved them to an abandoned electronics facility in Newberg. They wanted to produce more wine. Every business wants to grow—but it's a question of *how* you grow. People who love the quality of their wine want to be able to gain in reputation and hence become able to charge more per bottle. But the corporate strategy tends to be different— just ramp up the volume and exploit the preexisting reputation. It has been happening in the North coast since the '80s. Although the Argyle tasting room remained in Dundee, my office was moved to Newberg, which I found irritating. The company's values no longer reflected my passions about grapes and excellent wine. Decisions were being made far away where I could not see what was happening. After thirty years with Argyle, it was time to move on.

THE GENERATIONAL TRANSITION

ONE OF THE HARDEST THINGS for the Baby Boom generation of vineyard and winery owners to figure out was how to draw their children into the business. They followed different paths.

Cody Wright, Ken Wright's son, wanted to start his own business and not rest on his father's laurels. He made wine for a while at Ken Wright Cellars and later at Roco. I sold him grapes and delivered them to him there. He made a Holstein Vineyard label that got some attention. What a turn of events! Ken and I had gotten started together, and now I was selling grapes to his son. I did not fully appreciate the march of time until much later. Cody built his own wine-producing facility in Dundee. I knew of a party in Dundee who wanted to sell a vineyard, and I knew Cody had a partner who wanted to buy one. I put the parties together. Cody now runs a fully integrated operation, from vineyard to winery.

Jim Maresh, son of Martha and the late Fred Arterberry and grandson of Jim Maresh, the patriarch of Dundee, also started a wine brand. I sold him grapes too for a while, but his grandfather owned a comparable vineyard, and he started to source from there. In 2022, Jim had a Dundee Pinot Noir in the Top 100 Wines of the World in the Wine Spectator. He also is fully integrated.

In 2013, as my son, Jackson, was finishing college, he asked me if he could have a ton of grapes to make wine with a buddy in Corvallis, home of Oregon State University. But in 2014, he announced he wanted to start a wine business and would make wine at the Tori Mor winery if I would give him three tons of grapes. He took a day job with a vineyard management company to make ends meet. We bumped along like that for a while until it came time to sell the wine. It started with a tasting every now and then at my home. Naturally, people were impressed with the view and the wine. It was a direct-to-consumer model. But he was paying good money to the facility where he made wine, and the more wine he made the more he would have to pay. It was an impediment to growth. I wanted him to be successful.

So, the idea emerged: "Why don't we build a winery here?' He could use the money he was paying Tori Mor to pay for his own facility. Jackson did the leg work as far as permitting and design were concerned. We built it in 2018 out behind my house. Before long, he was charging his buddies to make wine there as well as making wine for other parties, such as three NBA basketball stars who wanted the prestige of having their own brands of wine. It served to diversify his income.

He and wife Ayla launched their own label called Granville, which was the first name of my great grandfather, Granville O'Dell. I obviously liked that. He was an original "back to the land" type in rural West Virginia.

I leased my vineyard to Jackson, which meant he was taking control of the management decisions about it. I had to let go. His perspective was different. I had a mantra, meanwhile, that I kept singing to myself . . . "Lead, follow, or let go!"

Another challenge was that the tastings at my house began to increase in number until it got to the point that it was becoming increasingly difficult to live there. Different tour groups from places like Minneapolis would come up to the house, maybe five of them a day. Jackson and Ayla were doing real business.

This is the heart of the problem involved in the generational transfer of wine assets. The young people want to move quickly to take over and show what they can do. But the older generation may not be quite ready to ride into the sunset.

Eventually, we sought professional help from a Certified Professional Accountant (CPA) who specialized in family business succession in the wine region. He proposed a solution. In 2021, Jackson and wife Ayla began to lease the entire premises, which allowed me to buy another house in Dundee and move there. Their business was growing, and I was happy for them and continued to enjoy visiting my property up on the hill.

Planting trees in front of my new house in 1987, first year of Argyle

Thirty years later, see how the trees have grown.

Looking south from Holstein Vineyards in fall

This picture shows the challenges vineyard managers have when making an observation about a block and extrapolating to the whole field based on ducking into the corner.

CHAPTER EIGHT:

Climate: The Big Question

WINE REGIONS AROUND THE WORLD are beginning to cope with climate change. According to a recent segment on CBS's 60 Minutes by Leslie Stahl, French winemakers are experimenting with different varieties to see which ones can perform best in hotter weather. It only stands to reason: growing any agricultural product depends on a cycle of flowering and ripening. Farmers have to understand how to match a specific plant species with specific weather patterns.

Beginning in 2014, the full impacts of climate change hit us in the eyes. Starting that year, we had above-average temperatures for several years. That allowed the grapes to reach full ripeness sooner and allowed for yields higher than we had been able to achieve in previous years. The flip side was that it forced us to compress the seasons, meaning we had to get more work done in a shorter period of time. But consistency improved, and Brand Oregon shined.

Things took a turn for the worse, climactically speaking. I was at the Oregon Coast for Labor Day in 2020. Before heading back up over the Coastal mountain range to Dundee, my friend and I went for a walk. As we walked, I got an alert on my phone about high winds and a fire warning back in Dundee. We drove up into the Coastal range and went through a little crossroads town called Otis that has a famous restaurant of the same name.

As we dropped down into the Willamette Valley, we noticed a high-level haze. It was the time of year when filbert growers prepared their fields for harvest, and they often kicked up dust. Usually, that was fairly localized. What we saw was more widespread. But it did not ring my alarm bell. I did not understand what was happening.

I was home alone at the vineyard that night and not paying attention to the phone. The whole valley had lit up. Otis had been closed down shortly after we passed through it, and the highway we had traveled on was closed because of a raging fire. We would have been trapped on the coast if we had waited much longer to make the return trip. Meanwhile, fires in the foothills of the Cascades on the other side of the Willamette Valley had lit up. I could look over to the Chehalem Mountains just four miles north and see flames and the red lights of fire trucks. It was a shock and awe campaign.

The next morning, we had smoke at a high level, and it was kept high by the strong winds from the east, which is an unusual weather event in and of itself. As the winds died down, however, the smoke dropped down on top of us. I had never seen anything like it and could not have imagined it could ever happen.

Immediately, concerns about the grapes being smoke-tainted emerged. Nobody had any experience with it. There are labs that can test for the molecules that cause it, but California, where most major labs were located, was having its own fires. There was a backlog so that no results could be obtained before harvest time. The advice went out to make wine in five gallon buckets and then taste it. That is where things stood. Some of the larger wineries did get test results, and news went out that buyers were canceling grape purchase deals. Or they said they would accept the grapes but if they were spoiled, they would not pay for them.

Air quality took a deep dive. Websites showed we had the worst air quality in the world.

The question remained, do you harvest or not? Some people abandoned their grapes. But my son, Jackson, harvested. After a month or so, the report from the lab was that the wine had smoke taint. Other wineries concurred. Domaine Drouhin did not make red wine that year. Months later, Jackson was able to sell the bulk wine to a larger-low price producer, and he got his investment out. I wondered about the rest of the industry. Never had we had a year where we got zero return. I pondered the future for growers that just sell grapes and do not make wine. It was a big hit. Combine the pandemic with a major natural disaster in the middle of harvest and you get a spin out. Drought is another threat that is obviously related to the fires.

On the opposite end of the threat spectrum are blasts of cold air from the Arctic. I remember in 1989 we had a high temperature of 50 °F on a Sunday in February. By the following Thursday, we had a low of 4 °F because of an Arctic air mass dropping out of Alberta, Canada. That was when we had own-rooted vines. If that happened today with grafted vines, it could be a disaster. Pinot Noir is not as cold hardy as rootstocks. If it got too cold, the plant could be damaged down to the graft union, in which case the vines would have to be grafted again.

Related but different from Arctic blasts in the winter are spring frosts. In 2022, upon nearing the finish line for this book, the Willamette Valley experienced an unprecedented frost on the nights of April 13 and 14. The week before, we had a couple of days in the mid-70s, which was unusual. This forced the buds developmentally into a vulnerable state and, as a result, the frost inflicted considerable damage. It took almost a month for the full extent of it to become obvious. Some of my vineyard manager buddies, who cover more ground than I do, think that at least 50 percent of the crop was damaged valley wide.

I talked with Alex Blosser. He was like a deer in the headlights. "After the smoke hit in 2020 and now this, what do we do?" he asked. Lower elevations were hit harder by the frost. I drove around thinking of the people I knew behind the vineyards I was passing. My heart broke for them. My vineyard seemed to be above the elevation threshold and was largely spared. I was happy for Jackson. "Better lucky than smart."

One last threat is that of exotic pests that are native to one part of the world but show up in Oregon. In the old days, plants had time to evolve with the pests and develop a tolerance or resistance to them through evolution. Pests learned that a good pest never kills its host. A balance developed. In the past two hundred years, that balance has been challenged by world travel and the global movement of products. If a plant was never exposed to a certain pest, it has not had an evolutionary opportunity to find ways to coexist. Powdery mildew is native to the eastern United States and has long been a problem in Europe and in the western United States. Now, mildew has developed resistance to the fungicides used to control it.

The current problem is called the Red Blotch, which is a virus that can delay maturity of the grapes and therefore reduce quality. Weeds from other parts of the world come in. Italian Thistle is the latest example. Japanese Beetle is a voracious insect that eats anything green. States have departments of agriculture whose job it is to control these pests, but they cannot keep up.

Does all this mean I am turning into a pessimist? No, and I will explain why. The Willamette Valley stands out among the Northwestern wine regions as having a lower risk. We are forty miles as the crow flies over the coastal range to the mighty Pacific Ocean and the winter rains it brings. Dundee's soils are deep, and they spend their winters soaking the rain up.

Another advantage we have is that only the warmest sites in Oregon have been planted with grapes. We did that because there was a risk in our cool climate of not being able to fully ripen the grapes. That meant that we did not go close to the Pacific Coast or go up too high in elevation in the coastal range where the weather is cooler. If climate change means more warming, Oregon has plenty of sites that would accommodate it. In the worst case scenario, the ideal sites of yesteryear could become less desirable, but it is hard to imagine them becoming unusable. So that is a strength.

Another reason I am confident is that the ownership structure of the Oregon wine industry has changed so fundamentally beyond anything the early pioneers could have imagined. Although I was never comfortable personally with distant corporate decision-makers, I have to acknowledge that the people who now own much of Oregon's vineyards and wineries have deeper pockets and better organizations than the pioneers ever had.

I don't think the big European companies will stop coming any time soon. If you were the owner or chief executive of a venerable wine business, aware of the potential climate impacts in Europe, what would you do? Land prices in Oregon are still a fraction of those in the most prestigious regions of the world. I think this is one of the lessons of Santa Margherita taking majority control of Rollin Soles' Roco Winery. More Italian companies are likely to come. That was the pattern from Burgundy. Drouhin was first, but ultimately his direct competitors had to establish a presence in Oregon as well. It was a question of being in position to compete in the United States, now the world's largest wine market. These are precisely the types of companies that have deep international expertise. And they are in it for the long term—they think in terms of generations of their families. "Look at recent history," Rollin Soles says. "Jadot, Henriot/Bouchard, Bollinger, and Santa Margherita represent the best sort of

Willamette Valley investments. Multi-generational Northern European winemakers believe in investing for the long term and are all focused on quality and strong sales/marketing of the wines in their portfolios. The young Willamette Valley wine region benefits from this sort of investment mindset. It also continues the European-Willamette cooperation and recognition we have all enjoyed over the past thirty years."

Regarding the increasing challenge of pests in a monoculture, I think we will be able to respond because of the collaboration we have established between growers and professors at Oregon State University, another example of the collaborative culture we have established. The problem in most wine regions of the world is that agriculture professors in their ivory towers have no idea what real-world conditions growers are facing. But in the early 2000s, I started what became the North Willamette Technical Group. It started with one professor and half a dozen growers. When Patty Skinkis started at OSU in 2007, we asked her to co-chair with myself. We took professors out to the vineyard and showed them real life problems. I have stepped back as co-chair but still go to meetings. Now there are thirty-five people exchanging ideas about solutions to exotic pests as well as other problems. I think this will be our secret weapon against any one pest gaining an upper hand.

We have repeatedly demonstrated our ability to adapt. We now possess the infrastructure and experience base to continue doing that. So all in all, I remain an optimist. I believe the future is bright for Brand Oregon. Véronique Drouhin also recognizes the challenges but largely agrees with my assessment, saying "The future of Oregon looks great." Here are her parting thoughts in response to my questions:

HAS THE AMERICAN KNOWLEDGE OF wine improved? "Wine knowledge in the United States used to be not so great. You had some well-known experts, but there were very few of them. Now it's unbelievable how many people know wine well and know wine from all over the world. The knowledge of wine has increased a lot. People have the culture. They know how to taste and how to appreciate the differences in wine. The sommeliers are very important and very well-educated. The younger generation has been traveling a lot, and the sommeliers have been traveling all over the world. And, of course, Pinot Noir has come to be known as a very refined, delicate, fruit-friendly wine."

What are your thoughts on climate change? "We have to acknowledge the challenges. We cannot deny them. We cannot deny that the weather is changing. For now, we see it more here in Burgundy than in Oregon. We have more violent episodes of frost or hail. In Oregon, we see it in the dates when we start picking grapes. It's a bit sooner than it has been in the past. We all should be aware that we need to be actors in the climate challenge and not just to do things to adapt but try to do things to prevent the change.

For example, when we ferment wine, a lot of carbon dioxide (CO_2) goes into the atmosphere. We know that's not good. We're not doing it yet at DDO, but I'm hoping to work on a way to trap that. We should be proactive. At the same time, we have to keep trying to make the best possible wines. We may have to work with different plants in Burgundy that bud later. Maybe we will have to embrace that in Oregon too someday."

Will more Europeans establish their presence in Oregon? "They have started to arrive, and I would not be surprised if some others will come. Oregon is very attractive. Here in Burgundy, it's very difficult to buy land for vineyards. It's extremely expensive, and the government regulates it heavily. In Oregon, there is still good land to buy, and it is affordable. California has its issues with drought, heat, and fires. If I were an investor, I would not invest in California. People in the wine industry often ask me about Oregon. I can feel that there is great interest. I definitely think we will see more people coming from outside the country.

"Others from Burgundy could come too. It has taken them a long time. I don't think they believed we would be successful. But we've been proving that Oregon is very serious. Of the thirty-four vintages since we started, all of them, except for 1997, are delicious and still lovely today. They are good for a long time and age well, which is great. Nobody knew that four decades ago."

What was your attitude about making money? "When we went to Dundee Hills, we knew we were not investing in a business that would be quickly profitable. It was about creating an exciting adventure. We believed in it. Of course, the idea was to create an estate that someday could live on its own. The Europeans, from the Old World, believe you should build something for the long term."

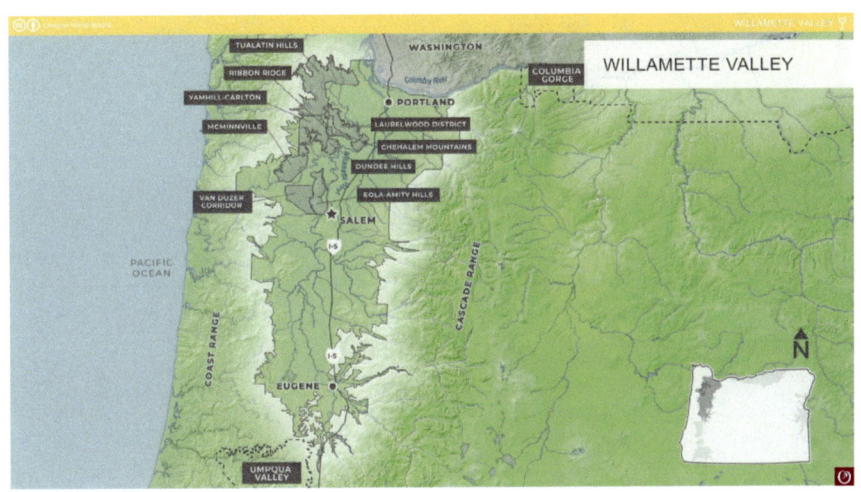

Map of Willamette Valley illustrates various American Viticulture Areas (AVA)

Acknowledgements

SPECIAL THANKS TO MY BROTHER, Bill Holstein. He and I were raised to chase our dreams, and because of that, we led almost polar opposite lives—me, living in Dundee, Oregon, for forty-two years and focusing on local issues and possessing a deep sense of place. Bill has lived abroad and currently lives in suburban New York. He has written extensively about a variety of global subjects. He has visited me regularly in Oregon, the first time in 1982 when he and my dad came to help me in the harvest. He has had snap shots ever since on a regular basis. In 2021, he visited and suggested I write a book because he could also see how much the Dundee region had changed. I provided him with two hundred pages of stream of consciousness commentary that included tons of minutia. He found the narrative arcs. His most recent book is entitled "A Grand Strategy: Countering China, Taming Technology, and Restoring the Media." For more about him, see his website: williamjholstein.com.

A note of gratitude is also in order to my partner in life, Om Sukheenai. She has been there for me during life's ups and downs and particularly supportive during this book effort when at times I lost focus on the next steps. She also helped make this happen. Thanks, Om.

Also, thanks go to Ken Wright and Veronique Drouhin, who gave considerable time in providing comment.

Finally, it would not have been the same without all of the mentors I had along the way in the educational, business, and wine sectors as well as my travel sponsors. My father Bill Holstein and Wendell Berry taught me the love of nature and the outdoors. They all combined to create a rich experience for me. I hope I have done the same for those who follow.